Ben Stacy Jerrik (Ed.)

Budge Budge I (Community Development Block)

Ben Stacy Jerrik (Ed.)

Budge Budge I (Community Development Block)

Alipore Sadar subdivision, South 24 Parganas district, Budge Budge, Balarampur, Budgebudge, Community development block in India

Part Press

Imprint

Permission is granted to copy, distribute and/or modify this document under the terms of the GNU Free Documentation License, Version 1.2 or any later version published by the Free Software Foundation; with no Invariant Sections, with the Front-Cover Texts, and with the Back- Cover Texts. A copy of the license is included in the section entitled "GNU Free Documentation License".

All parts of this book are extracted from Wikipedia, the free encyclopedia (www.wikipedia.org).

You can get detailed informations about the authors of this collection of articles at the end of this book. The editors (Ed.) of this book are no authors. They have not modified or extended the original texts.

Pictures published in this book can be under different licences than the GNU Free Documentation License. You can get detailed informations about the authors and licences of pictures at the end of this book.

The content of this book was generated collaboratively by volunteers. Please be advised that nothing found here has necessarily been reviewed by people with the expertise required to provide you with complete, accurate or reliable information. Some information in this book maybe misleading or wrong. The Publisher does not guarantee the validity of the information found here. If you need specific advice (f.e. in fields of medical, legal, financial, or risk management questions) please contact a professional who is licensed or knowledgeable in that area.

Any brand names and product names mentioned in this book are subject to trademark, brand or patent protection and are trademarks or registered trademarks of their respective holders. The use of brand names, product names, common names, trade names, product descriptions etc. even without a particular marking in this works is in no way to be construed to mean that such names may be regarded as unrestricted in respect of trademark and brand protection legislation and could thus be used by anyone.

Cover image: www.ingimage.com
Concerning the licence of the cover image please contact ingimage.

Publisher:
Part Press is a trademark of
International Book Market Service Ltd., 17 Rue Meldrum, Beau Bassin, 1713-01 Mauritius
Email: info@bookmarketservice.com
Website: www.bookmarketservice.com

Published in 2012

Printed in: U.S.A., U.K., Germany. This book was not produced in Mauritius.

ISBN: 978-613-6-28323-4

Contents

Budge_Budge_I_(community_development_block)

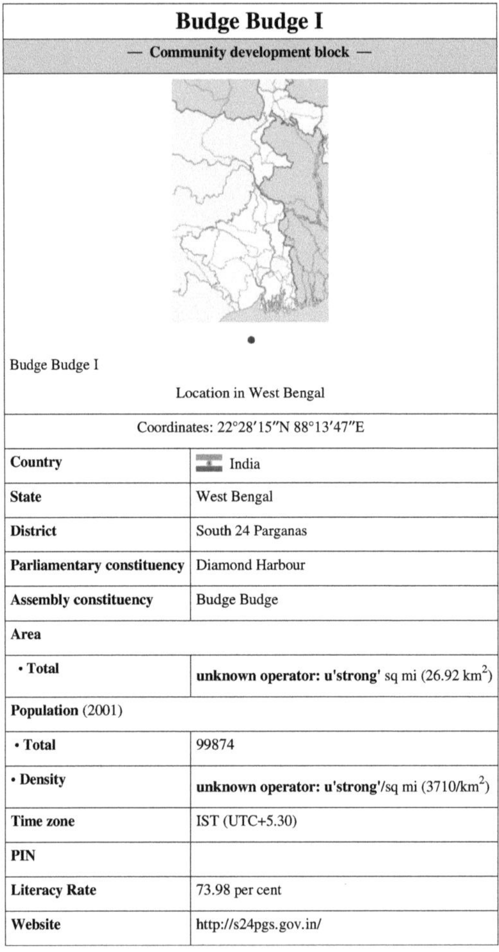

<table>
<tr><td colspan="2" align="center">Budge Budge I</td></tr>
<tr><td colspan="2" align="center">— Community development block —</td></tr>
<tr><td colspan="2" align="center">Budge Budge I

Location in West Bengal</td></tr>
<tr><td colspan="2" align="center">Coordinates: 22°28′15″N 88°13′47″E</td></tr>
<tr><td>Country</td><td>India</td></tr>
<tr><td>State</td><td>West Bengal</td></tr>
<tr><td>District</td><td>South 24 Parganas</td></tr>
<tr><td>Parliamentary constituency</td><td>Diamond Harbour</td></tr>
<tr><td>Assembly constituency</td><td>Budge Budge</td></tr>
<tr><td>Area</td><td></td></tr>
<tr><td>• Total</td><td>unknown operator: u'strong' sq mi (26.92 km^2)</td></tr>
<tr><td>Population (2001)</td><td></td></tr>
<tr><td>• Total</td><td>99874</td></tr>
<tr><td>• Density</td><td>unknown operator: u'strong'/sq mi (3710/km^2)</td></tr>
<tr><td>Time zone</td><td>IST (UTC+5.30)</td></tr>
<tr><td>PIN</td><td></td></tr>
<tr><td>Literacy Rate</td><td>73.98 per cent</td></tr>
<tr><td>Website</td><td>http://s24pgs.gov.in/</td></tr>
</table>

Budge Budge I (community development block) (Bengali: বজ বজ I সমষ্টি উন্নয়ন ব্লক) is an administrative division in Alipore Sadar subdivision of South 24 Parganas district in the Indian state of West Bengal. Budge Budge police station serves this block. Headquarters of this block is at Purba Nischintapur. Balarampur, Uttar Raypur and Birlapur are urban areas in this block.[1] [2]

Geography

Chingripota, a constituent panchayat of Budge Budge I block, is located at 22°28′15″N 88°13′47″E.

Budge Budge I community development block has an area of 26.92 km^2.[2]

Gram panchayats

Gram panchayats of Budge Budge I block/panchayat samiti are: Buita, Chingripota, Mayapur, Nischintapur, Rajibpur and Uttar Raipur. [3]

Demographics

As per 2001 census, Budge Budge I block had a total population of 99,874, out of which 98,027 were males and 92,589 were females. Budge Budge I block registered a population growth of -16.38 per cent during the 1991-2001 decade. Decadal growth for South 24 Parganas district was 20.89 per cent.[2] Decadal growth in West Bengal was 17.84 per cent.[4]

Scheduled castes at 23,829 formed around one-fourth the population. Scheduled tribes numbered 660.[5]

Literacy

As per 2001 census, Budge Budge I block had a total literacy of 73.98 per cent for the 6+ age group. While male literacy was 81.36 per cent female literacy was 64.82 per cent. South 24 Parganas district had a total literacy of 69.45 per cent, male literacy being 79.19 per cent and female literacy being 59.01 per cent.[6]

References

[1] "Contact details of Block Development Officers" (http://wbprd.gov.in/html/asp/bdo_contact.asp?cd=EJ). *South 24 Parganas district*. West Bengal Government. . Retrieved 2011-09-24.

[2] "Provisional population totals, West Bengal, Table 4, South Twentyfour Parganas District (18)" (http://web.cmc.net.in/wbcensus/ DataTables/02/Table4_18.htm). *Census of India 2001*. Census Commission of India. . Retrieved 2011-09-24.

[3] "No. 229 (Sanction)-PN/P/II/1G-5/2005(Pt.II) dated 02.02.09" (http://wbprd.nic.in/html/asp/writereaddata/Notifications/01070024. doc). *Annexure A of 322 (Sanction)-RD/CCA/BRGF/IC-02/08; Date: 27.03.2009 – Statement of Release of Development Grant under BRGF*. Government of West Bengal - Department of Panchayats & Rural Development. . Retrieved 2011-09-24.

[4] "Provisional Population Totals, West Bengal. Table 4" (http://web.cmc.net.in/wbcensus/DataTables/02/FrameTable4_1.htm). *Census of India 2001*. Census Commission of India. . Retrieved 2011-09-23.

[5] "TRU for all Districts (SC & ST and Total)" (http://web.cmc.net.in/wbcensus/HouseListingF/SCST/All_distSCST(TRU1)17.htm). *Census 2001*. Census Commission of India. . Retrieved 2011-09-24.

[6] "Provisional population totals, West Bengal, Table 5, South Twentyfour Parganas District" (http://web.cmc.net.in/wbcensus/DataTables/ 02/Table5_18.htm). *Census of India 2001*. Census Commission of India. . Retrieved 2011-09-24.

Alipore_Sadar_subdivision

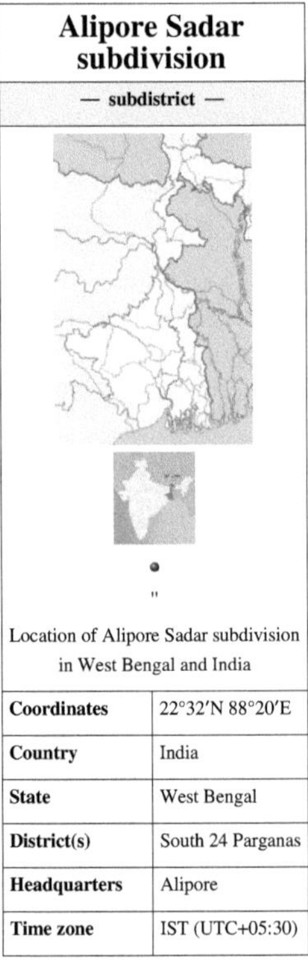

Alipore Sadar subdivision

— subdistrict —

Location of Alipore Sadar subdivision in West Bengal and India

Coordinates	22°32′N 88°20′E
Country	India
State	West Bengal
District(s)	South 24 Parganas
Headquarters	Alipore
Time zone	IST (UTC+05:30)

Alipore Sadar subdivision (Bengali: আলিপুর সদর মহকুমা), is a subdivision of the South 24 Parganas district in the state of West Bengal, India. It consists of three municipalities (Budge Budge, Pujali and Maheshtala) and five community development blocks: Bishnupur–I, Bishnupur–II, Budge Budge–I, Budge Budge–II and Thakurpukur Maheshtala. The five blocks contain ten census towns and 45 gram panchayats. Apart from these, a large parts of Calcutta Municipal Corporation fall in this Apipore sadar subdivision. The subdivision has its headquarters at Alipore.

Area

Apart from the municipalities of Budge Budge, Pujali and Maheshtala, the subdivision contains ten census towns and rural areas of 45 gram panchayats under five community development blocks: Bishnupur–I, Bishnupur–II, Budge Budge–I, Budge Budge–II and Thakurpukur Maheshtala.[1] The ten census towns are: Bishnupur, Kanyanagar, Amtala, Chak Enayetnagar, Balarampur, Uttar Raypur, Birlapur, Chak Kashipur, Bowali and Chata Kalikapur.[2]

Blocks

Bishnupur–I block

Rural area under Bishnupur–I block consists of 11 gram panchayats, viz. Amgachhia, Dakshin Gauripur Chakdhir, Kulerdari, Raskhali, Andhar Manik, Panakua, Bhandaria Kastekumari, Julpia, Paschim Bishnupur, Keoradanga and Purba Bishnupur.[1] Urban area under this block consists of two census towns: Bishnupur and Kanyanagar.[2] Bishnupur police station serves this block.[3] Headquarters of this block is in Bishnupur.[4]

Bishnupur–II block

Rural area under Bishnupur–II block consists of 11 gram panchayats, viz. Bakhrahat, Kanganberia, Patharberia Jaychandipur, Chak Enayetnagar, Khagramuri, Chandi, Maukhali, Ramkrishnapur Borhanpur, Gobindapur Kalicharanpur, Nahajari and Panchanan.[1] Urban area under this block consists of two census towns: Amtala and Chak Enayetnagar.[2] Bishnupur police station serves this block.[3] Headquarters of this block is in Bakhrahat.[4]

Budge Budge–I block

Rural area under Budge Budge–I block consists of six gram panchayats, viz. Buta, Mayapur, Rajibpur, Chingripota, Nischintapur and Uttar Raipur.[1] Urban area under this block consists of three census towns: Balarampur, Uttar Raypur and Birlapur.[2] Budge Budge police station serves this block.[3] Headquarters of this block is in Purba Nishchintapur.[4]

Budge Budge–II block

Rural area under Budge Budge–II block consists of 11 gram panchayats, viz. Burul, Gaja Poyali, Naskarpur, Satgachhia, Chakmanik, Kamrabad, North Baoyali, South Baoyali, Dongaria Raipur, Kasipur Alampur and Rania.[1] Urban area under this block consists of two census towns: Chak Kashipur and Bowali.[2] Nadakhali police station serves this block.[3] Headquarters of this block is in Dongaria.[4]

Thakurpukur Maheshtala block

Rural area under Thakurpukur Maheshtala block consists of six gram panchayats, viz. Asuti–I, Joka–I, Asuti–II, Joka–II, Chatta and Raspunja.[1] Urban area under this block consists of one census town: Chata Kalikapur.[2] Thakurpukur, Maheshtala and Bishnupur police stations serve this block.[3] Headquarters of this block is in Biran Ray Road (East).[4]

Legislative segments

As per order of the Delimitation Commission in respect of the delimitation of constituencies in West Bengal, the area under Bishnupur–II block along with seven gram panchayats under the Budge Budge–II block, viz. Burul, Chakmanik, Gaja Poyali, Kamrabad, Naskarpur, Rania and Satgachhia, will form the **Satgachhia assembly constituency** of West Bengal. The other four gram panchayats under the Bishnupur–II block, the Budge Budge municipality, the Pujali municipality and the area under the Budge Budge–I block will together form the **Budge Budge assembly constituency**. The gram panchayats of Asuti–I, Asuti–II, Chatta and Raspunja under the

Thakurpukur Maheshtala block and the area under the Bishnupur–I block will together form the **Bishnupur assembly constituency**. Wards 66,67, 91, 92, 107 and 108 of Kolkata municipal corporation will together form the **Kasba assembly constituency**. Wards 96, 99, 101–106, 109 and 110 of Kolkata municipal corporation will form the **Jadavpur assembly constituency**. Wards 94, 95, 97, 98, 100, 111–114 of Kolkata municipal corporation will form the **Tollygunj assembly constituency**. Wards 118, 119, 125–132 of Kolkata municipal corporation will form the **Behala Paschim assembly constituency**. The **Behala Purba assembly constituency** will cover the area under wards 115–117 and 120–124 of the Kolkata municipal corporation, and the gram panchayats of Joka–I and Joka–II under the Thakurpukur Maheshtala block. Wards 8, 11–35 of the Maheshtala municipality will form the **Maheshtala assembly constituency**. Wards 1–7, 9, 10 of the Maheshtala municipality and wards 136–141 of the Kolkata municipal corporation will together form the **Metiaburuz assembly constituency**. The Bishnupur constituency will be reserved for Scheduled castes (SC) candidates. The constituencies of Satgachhia, Bishnupur, Maheshtala, Budge Budge and Metiaburuz will be assembly segments of the **Diamond Harbour (Lok Sabha constituency)**. The constituencies of Jadavpur and Tollyganj will be segments of the **Jadavpur (Lok Sabha constituency)**. The constituencies of Kasba, Behala Paschim and Behala Purba will be assembly segments of the **Kolkata Dakshin (Lok Sabha constituency)**.[5]

References

[1] "Directory of District, Sub division, Panchayat Samiti/ Block and Gram Panchayats in West Bengal, March 2008" (http://wbdemo5.nic.in/ writereaddata/Directoryof_District_Block_GPs(RevisedMarch-2008).doc). *West Bengal*. National Informatics Centre, India. 2008-03-19. . Retrieved 2008-12-24.

[2] "District Wise List of Statutory Towns(Municipal Corporation,Municipality,Notified Area and Cantonment Board) , Census Towns and Outgrowths, West Bengal, 2001" (http://web.cmc.net.in/wbcensus/DataTables/01/Table-3.htm). Census of India, Directorate of Census Operations, West Bengal. . Retrieved 2008-12-24.

[3] "List of Districts/C.D.Blocks/ Police Stations with Code No., Number of G.Ps and Number of Mouzas" (http://web.cmc.net.in/wbcensus/ DataTables/01/FrameTable2_8.htm). Census of India, Directorate of Census Operations, West Bengal. . Retrieved 2008-12-24.

[4] "Contact details of Block Development Officers" (http://wbdemo5.nic.in/html/asp/bdo_contact.asp?cd=EJ). *South 24 Parganas district*. Panchayats and Rural Development Department, Government of West Bengal. . Retrieved 2008-12-27.

[5] "Press Note, Delimitation Commission" (http://www.wbgov.com/e-gov/English/DELIMITATION.pdf) (PDF). *Assembly Constituencies in West Bengal*. Delimitation Commission. pp. 12–14, 24. . Retrieved 2009-01-17.

South_24_Parganas_district

<table>
<tr><td colspan="2" align="center">South 24 Parganas district</td></tr>
<tr><td colspan="2">

Location of South 24 Parganas district in West Bengal</td></tr>
<tr><td>State</td><td>West Bengal, India</td></tr>
<tr><td>Administrative division</td><td>Presidency</td></tr>
<tr><td>Headquarters</td><td>Alipore</td></tr>
<tr><td>Area</td><td>9960 km^2 (unknown operator: u'strong' sq mi)</td></tr>
<tr><td>Population</td><td>81,53,176 (2011)</td></tr>
<tr><td>Population density</td><td>819 /km^2 (unknown operator: u'strong' /sq mi)</td></tr>
<tr><td>Urban population</td><td>1,089,730</td></tr>
<tr><td>Literacy</td><td>78.57 per cent[1]</td></tr>
<tr><td>Sex ratio</td><td>937</td></tr>
<tr><td>Lok Sabha constituencies</td><td>Jaynagar, Maturapur, Diamond Harbour, Jadavpur, Kolkata Dakshin</td></tr>
<tr><td>Assembly seats</td><td>Gosaba, Basanti, Kultali, Patharpratima, Kakdwip, Sagar, Kulpi, Raidighi, Mandirbazar, Jaynagar, Baruipur Purba, Canning Paschim, Canning Purba, Baruipur Paschim, Magrahat Purba, Magrahat Paschim, Diamond Harbour, Falta, Satgachia, Bishnupur, Sonarpur Dakshin, Bhangar, Kasba, Jadavpur, Sonarpur Uttar, Tollyganj, Behala Purba, Behala Paschim, Maheshtala, Budge Budge, Metiaburuz</td></tr>
<tr><td>Major highways</td><td>NH 117</td></tr>
<tr><td>Average annual precipitation</td><td>1750 mm</td></tr>
<tr><td colspan="2" align="center">Official website [2]</td></tr>
</table>

South 24 Parganas district (Bengali: দক্ষিণ চব্বিশ পরগণা জেলা) is an important district of West Bengal State with its district headquarters in Alipore. It has the urban fringe of Kolkata on one side and the remote riverine villages in the Sundarbans.

It is the sixth most populous district in India (out of 640)[3]

History

Once the capital of Raja Bikramaditya and Maharaja Pratapaditya was at Dhumghat. Later it was transferred to Ishwaripur (Originated from the name Jeshoreshwaripur). Maharaja Pratapaditya declared independence of South Bengal (Jessore, Khulna in north, Sundarbans, Bay of Bengal in South, Barisal in east and River Ganges in west) against the Mughal Empire of India.

Jashoreshwari Kali Temple (built by Pratapaditya), Chanda Bhairab Mandir at Ishwaripur (a triangular temple, built during the Sena period), Five domed Tenga Mosque at Banshipur (Mughal period), two big and four small domed Hammankhana (constructed by Pratapaditya) at Bangshipur, Govinda Dev Temple at Gopalpur (built by Basanta Roy, uncle of Maharaja Pratapaditya in 1593), Jahajghata Port (Khanpur). Pratapaditya king of Jessore and one of the bara-bhuiyans of Bengal. Pratapaditya fought against the Mughal imperial army during its inroad into Bengal in the early 17th century. His father Shrihari (Shridhar), a Kayastha, was an influential officer in the service of daud khan karrani. On the fall of Daud he fled away with the government treasure in his custody. He then set up a kingdom for himself in the marshy land to the extreme south of Khulna district (1574) and took the title of Maharaja. Pratapaditya succeeded to the kingship in 1574. The baharistan and the travel diary of Abdul Latif and the contemporary European writers, all testify to the personal ability of Pratapaditya, his political pre-eminence, material resources and martial strength, particularly in war-boats. His territories covered the greater part of what is now included in the greater Jessore, Khulna and Barisal districts. He established his capital at Dhumghat, a strategic position at the confluence of the Jamuna and Ichhamati.

Among the Bengal zamindars Pratapaditya was the first to send his envoy to Islam Khan Chisti with a large gift to win the favour of the Mughals, and then tendered personal submission to the Subahdar (1609). He promised military assistance and personal service in the Mughal campaign against musa khan, a pledge that he did not keep. To punish Pratapaditya for his disloyalty as a vassal and to subjugate his territory, a large expedition was launched under the command of Ghiyas Khan, which soon reached a place named Salka, near the confluence of the Jamuna and Ichhamati (1611). Pratapaditya equipped a strong army and a fleet and placed them under expert officers including Feringis, Afghans and Pathans. His eldest son Udayaditya made a big fort at Salka with natural barriers on three sides rendering it almost impregnable. In battle the Jessore fleet gained an initial advantage. But the imperial army cut off the Jessore fleet, made a breach in its ranks and broke its unity and discipline. In the melee that followed, the admiral Khwaja Kamal was killed. Udayaditya lost heart and hastily fled to his father, narrowly escaping capture. Jamal Khan evacuated the fort and followed Udayaditya.

Pratapaditya prepared himself to fight a second time from a new base near the confluence of Kagarghat canal and the Jamuna. He made a big fort at a strategic point and gathered all his available forces there. The imperialists began the battle by an attack on the Jessore fleet (Jan 1612) and compelled it to seek shelter beneath the fort. But their further advance was checked by the heavy cannonade of the Jessore artillery. A sudden attack of the imperialists completely defeated the Jessore fleet and they fell upon the fort with the elephants in front, thereby compelling Pratapaditya to evacuate the fort and retreat.

The second defeat sealed the fate of Pratapaditya. At Kagarghat he tendered submission to Ghiyas Khan, who personally escorted Pratapaditya to Islam Khan at Dhaka. The Jessore king was put in chains and his kingdom was annexed. Pratapaditya was kept confined at Dhaka. No authentic information is available regarding his last days. Probably he died at Benares on his way to Delhi, as a prisoner.[4]

Economy

Agriculture, Industry and Pisciculture are all at their peak in the district. In west side of this district situated Falta Special Economic Zone.Various types of industry is situated in this Economic Zone.

In 2006 the Ministry of Panchayati Raj named South 24 Parganas one of the country's 250 most backward districts (out of a total of 640).[5] It is one of the eleven districts in West Bengal currently receiving funds from the Backward Regions Grant Fund Programme (BRGF).[5]

Divisions

Administrative subdivisions

The district comprises five subdivisions: Baruipur, Canning, Diamond Harbour, Kakdwip and Alipore Sadar. Baruipur subdivision consists of three municipalities (Baruipur, Rajpur Sonarpur and Jaynagar Mazilpur) and seven community development blocks: Baruipur, Bhangar–I, Bhangar–II, Jaynagar–I, Jaynagar–II, Kultali and Sonarpur. Canning subdivision consists of four community development blocks: Basanti, Canning–I, Canning–II and Gosaba. Diamond Harbour subdivision consists of Diamond Harbour municipality and nine community development blocks: Diamond Harbour–I, Diamond Harbour–II, Falta, Kulpi, Magrahat–I, Magrahat–II, Mandirbazar, Mathurapur–I and Mathurapur–II. Kakdwip subdivision consists of four community development blocks: Kakdwip, Namkhana, Patharpratima and Sagar. Alipore Sadar subdivision consists of three municipalities (Budge Budge, Pujali and Maheshtala) and five community development blocks: Bishnupur–I, Bishnupur–II, Budge Budge–I, Budge Budge–II and Thakurpukur Maheshtala.[6]

Alipore is the district headquarters. There are 33 police stations, 29 development blocks, 7 municipalities and 312 gram panchayats in this district.[6] [7] The Sunderbans area is covered by thirteen CD blocks, viz. Sagar, Namkhana, Kakdwip, Patharpratima, Kultali, Mathurapur–I, Mathurapur–II, Jaynagar–I, Jaynagar–II, Canning–I, Canning–II, Basanti and Gosaba.[7] This district contains 37 islands.[7]

Other than municipality area, each subdivision contains community development blocks which in turn are divided into rural areas and census towns. In total there are 21 urban units: 7 municipalities and 14 census towns.[7] [8]

Baruipur subdivision

- Three municipalities: Baruipur, Rajpur Sonarpur and Jaynagar Mazilpur.[7]
- Baruipur community development block consists of rural areas only with 19 gram panchayats.
- Bhangar–I community development block consists of rural areas with 9 gram panchayats and one census town: Bhangar Raghunathpur.
- Bhangar–II community development block consists of rural areas only with 10 gram panchayats.
- Jaynagar–I community development block consists of rural areas with 12 gram panchayats and one census town: Uttar Durgapur.
- Jaynagar–II community development block consists of rural areas only with 10 gram panchayats.
- Kultali community development block consists of rural areas only with 9 gram panchayats.
- Sonarpur community development block consists of rural areas only with 11 gram panchayats.

Canning subdivision

- Basanti community development block consists of rural areas only with 13 gram panchayats.
- Canning–I community development block consists of rural areas only with 10 gram panchayats.
- Canning–II community development block consists of rural areas only with 9 gram panchayats.
- Gosaba community development block consists of rural areas only with 14 gram panchayats.

Diamond Harbour subdivision

- One municipality: Diamond Harbour
- List of development blocks (9):

Development Block	Gram Panchayats	Remarks
Diamond Harbour I	8	
Diamond Harbour II	8	
Falta	13	
Kulpi	14	
Magrahat I	11	
Magrahat II	14	Also has two census towns - Uttar Kalas and Bilandapur
Mandirbazar	10	
Mathurapur II	10	
Mathurapur II	11	

Kakdwip subdivision

- Kakdwip community development block consists of rural areas only with 11 gram panchayats.
- Namkhana community development block consists of rural areas only with 7 gram panchayats.
- Patharpratima community development block consists of rural areas only with 15 gram panchayats.
- Sagar community development block consists of rural areas only with 9 gram panchayats.

Alipore Sadar subdivision

- Three municipalities: Budge Budge, Pujali and Maheshtala.
- Bishnupur–I community development block consists of rural areas with 11 gram panchayats and two census towns: Bishnupur and Kanyanagar.
- Bishnupur–II community development block consists of rural areas with 11 gram panchayats and two census towns: Amtala and Chak Enayetnagar.
- Budge Budge–I community development block consists of rural areas with 6 gram panchayats and three census towns: Balarampur, Uttar Raypur and Birlapur.
- Budge Budge–II community development block consists of rural areas with 11 gram panchayats and two census towns: Chak Kashipur and Bowali.
- Thakurpukur Maheshtala community development block consists of rural areas with 6 gram panchayats and one census town: Chata Kalikapur.

Parliamentary constituencies

The district has 5 parliament constituencies :[9]

- 19-Joynagar (SC) Parliamentary COnstituency
- 20-Mathurapur (SC) Parliamentary COnstituency
- 21-Diamond Harbour Parliamentary Constituency
- 22-Jadavpur Parliamentary Constituency
- 23-Kolkata Dakshin Parliamentary Constituency

Assembly constituencies

The district is divided into 32 assembly constituencies:[10]

1. Gosaba (SC) (assembly constituency no. 100),
2. Basanti (SC) (assembly constituency no. 101),
3. Kultali (SC) (assembly constituency no. 102),
4. Jaynagar (assembly constituency no. 103),
5. Baruipur (assembly constituency no. 104),
6. Canning West (SC) (assembly constituency no. 105),
7. Canning East (assembly constituency no. 106),
8. Bhangar (assembly constituency no. 107),
9. Jadavpur (assembly constituency no. 108),
10. Sonarpur (SC) (assembly constituency no. 109),
11. Bishnupur East (SC) (assembly constituency no. 110),
12. Bishnupur West (assembly constituency no. 111),
13. Behala East (assembly constituency no. 112),
14. Behala West (assembly constituency no. 113),
15. Garden Reach (assembly constituency no. 114),
16. Maheshtala (assembly constituency no. 115),
17. Budge Budge (assembly constituency no. 116),
18. Satgachia (assembly constituency no. 117),
19. Falta (assembly constituency no. 118),
20. Diamond Harbour (assembly constituency no. 119),
21. Magrahat West (assembly constituency no. 120),
22. Magrahat East (SC) (assembly constituency no. 121),
23. Mandirbazar (SC) (assembly constituency no. 122),
24. Mathurapur (assembly constituency no. 123),
25. Kulpi (SC) (assembly constituency no. 124),
26. Patharpratima (assembly constituency no. 125),
27. Kakdwip (assembly constituency no. 126),
28. Sagar (assembly constituency no. 127),
29. Kabitirtha (assembly constituency no. 147),
30. Alipore (assembly constituency no. 148),
31. Tollygunge (assembly constituency no. 150) and
32. Dhakuria (assembly constituency no. 151).

Gosaba, Basanti, Kultali, Canning West, Sonarpur, Bishnupur East, Magrahat East, Mandirbazar and Kulpi constituencies are reserved for Scheduled Castes (SC) candidates. Along with one assembly constituency from North 24 Parganas district, Gosaba, Basanti, Kultali, Jaynagar, Canning West and Canning East assembly constituencies form the Joynagar (Lok Sabha constituency), which is reserved for Scheduled Castes (SC). Baruipur, Jadavpur,

Bishnupur East, Behala East, Behala West, Magrahat West and Kabitirtha constituencies form the Jadavpur (Lok Sabha constituency). Bishnupur West, Garden Reach, Maheshtala, Budge Budge, Satgachia, Falta and Diamond Harbour constituencies form the Diamond Harbour (Lok Sabha constituency). Magrahat East, Mandirbazar, Mathurapur, Kulpi, Patharpratima, Kakdwip and Sagar constituencies form the Mathurapur (Lok Sabha constituency), which is reserved for Scheduled Castes (SC). Along with six assembly segments from North 24 Parganas district, Bhangar assembly constituency forms the Basirhat (Lok Sabha constituency). Along with three assembly constituencies from Kolkata district, Sonarpur, Alipore, Tollygunge and Dhakuria form the Calcutta South (Lok Sabha constituency).

Impact of delimitation of constituencies

As per order of the Delimitation Commission in respect of the delimitation of constituencies in the West Bengal, the district will be divided into 31 assembly constituencies:[11]

1. Gosaba (SC) (assembly constituency no. 127),
2. Basanti (SC) (assembly constituency no. 128),
3. Kultali (SC) (assembly constituency no. 129),
4. Patharpratima (assembly constituency no. 130),
5. Kakdwip (assembly constituency no. 131),
6. Sagar (assembly constituency no. 132),
7. Kulpi (assembly constituency no. 133),
8. Raidighi (assembly constituency no. 134),
9. Mandirbazar (SC) (assembly constituency no. 135),
10. Jaynagar (SC) (assembly constituency no. 136),
11. Baruipur Purba (SC) (assembly constituency no. 137),
12. Canning Paschim (SC) (assembly constituency no. 138),
13. Canning Purba (assembly constituency no. 139),
14. Baruipur Paschim (assembly constituency no. 140),
15. Magrahat Purba (SC) (assembly constituency no. 142),
16. Magrahat Paschim (assembly constituency no. 143),
17. Diamond Harbour (assembly constituency no. 144),
18. Falta (assembly constituency no. 118),
19. Satgachhia (assembly constituency no. 117),
20. Bishnupur (SC) (assembly constituency no. 110),
21. Sonarpur Dakshin (assembly constituency no. 109),
22. Bhangar (assembly constituency no. 107),
23. Kasba (assembly constituency no. 107),
24. Jadavpur (assembly constituency no. 108),
25. Sonarpur Uttar (assembly constituency no. 109),
26. Tollyganj (assembly constituency no. 150) and
27. Behala Purba (assembly constituency no. 112),
28. Behala Paschim (assembly constituency no. 113),
29. Maheshtala (assembly constituency no. 115),
30. Budge Budge (assembly constituency no. 116),
31. Metiaburuz (assembly constituency no. 147),

Gosaba, Basanti, Kultali, Mandirbazar, Jaynagar, Baruipur East, Magrahat East and Bishnupur constituencies will be reserved for Scheduled Castes (SC) candidates. Gosaba, Basanti, Kultali, Jaynagar, Canning West, Canning East and Magrahat East assembly constituencies will form the Jaynagar (Lok Sabha constituency), which will be reserved for Scheduled Castes (SC). Patharpratima, Kakdwip, Sagar, Kulpi, Raidighi, Mandirbazar and Magrahat West constituencies will form the Mathurapur (Lok Sabha constituency), which will be reserved for Scheduled Castes (SC). Diamond Harbour, Falta, Satgachia, Bishnupur, Maheshtala, Budge Budge and Metiaburuz constituencies will form the Diamond Harbour (Lok Sabha constituency). Baruipur East, Baruipur West, Sonarpur South, Bhangar, Jadavpur, Sonarpur North and Tollyganj constituencies will form the Jadavpur (Lok Sabha constituency). Along with six assembly segments from North 24 Parganas district, Bhangar assembly constituency forms the Basirhat (Lok Sabha constituency). Along with four assembly constituencies from Kolkata district, Kasba, Behala East and Behala West will form the Calcutta South (Lok Sabha constituency).

Demographics

According to the 2011 census South 24 Parganas district has a population of 8,153,176,[3] roughly equal to the nation of Honduras[12] or the US state of Virginia.[13] This gives it a ranking of 6th in India (out of a total of 640).[3] The district has a population density of 819 inhabitants per square kilometre (**unknown operator: u'strong'** /sq mi) .[3] Its population growth rate over the decade 2001-2011 was 18.05 %.[3] South Twenty Four Parganas has a sex ratio of 949 females for every 1000 males,[3] and a literacy rate of 78.57 %.[3]

Flora and fauna

In 1984 South 24 Parganas district became home to Sundarbans National Park, which has an area of 1330 km^2 (**unknown operator: u'strong'** sq mi).[14] It shares the park with North 24 Parganas district. It is also home to four wildlife sanctuaries: Haliday Island, Lothian Island, Narendrapur, and Sajnekhali.[14]

Sundarbans, formerly Sunderbunds, is a vast tract of forest and saltwater swamp forming the lower part of the Ganges Delta, extending about 160 miles (260 km) along the Bay of Bengal from the Hooghly River Estuary (India) to the Meghna River Estuary in Bangladesh. The whole tract reaches inland for 60–80 miles (100–130 km).

A network of estuaries, tidal rivers, and creeks intersected by numerous channels, it encloses flat, marshy islands covered with dense forests. The name Sundarbans is perhaps derived from the term meaning "forest of sundari," a reference to the large mangrove tree that provides valuable fuel. Along the coast the forest passes into a mangrove swamp; the southern region, with numerous wild animals and crocodile-infested estuaries, is virtually uninhabited. It is one of the last preserves of the Bengal tiger and the site of a tiger preservation project. The cultivated northern area yields rice, sugarcane, timber, and betel nuts.

The region is also famous for some commonly domesticated livestock breeds which includes the Garole breed of sheep and Chinae hans or Muscuovy ducks, the Garole sheep is considered as the progenator of the Booroola merino sheep and is noted for its prolific character. However, the wool of the sheep which can be a valuable natural asset does not find any use among the natives. Bakkhali beach resort located on one of the islands jutting out into the Bay of Bengal is gaining in popularity, with improvements in transport links with Kolkata.The woodlands and swamps of Sundarbans provide habitat to a number of species. Sundarbans is the homeland for Bengal Tiger, one of the most sought creatures around. The locality has a site for the preservation of tigers which is quite frequently visited by tourists and environmentalists. The estuaries in the region are infested with crocodiles. The area has been declared as world heritage site by the UNESCO. Boat tours are provided at many places in the region.

References

[1] "District-specific Literates and Literacy Rates, 2001" (http://www.educationforallinindia.com/page157.html). Registrar General, India, Ministry of Home Affairs. . Retrieved 2010-10-10.

[2] http://s24pgs.gov.in/

[3] "District Census 2011" (http://www.census2011.co.in/district.php). Census2011.co.in. 2011. . Retrieved 2011-09-30.

[4] Muazzam Hussain Khan (Banglapedia)

[5] Ministry of Panchayati Raj (September 8, 2009). "A Note on the Backward Regions Grant Fund Programme" (http://www.nird.org.in/ brgf/doc/brgf_BackgroundNote.pdf). National Institute of Rural Development. . Retrieved September 27, 2011.

[6] (RevisedMarch-2008).doc "Directory of District, Sub division, Panchayat Samiti/ Block and Gram Panchayats in West Bengal, March 2008" (http://wbdemo5.nic.in/writereaddata/Directoryof_District_Block_GPs). *West Bengal*. National Informatics Centre, India. 2008-03-19. (RevisedMarch-2008).doc. Retrieved 2008-12-03.

[7] "District Profile" (http://s24pgs.gov.in/main_Dist_Profile.htm). Official website of South 24 Parganas district. . Retrieved 2008-12-03.

[8] "Population, Decadal Growth Rate, Density and General Sex Ratio by Residence and Sex, West Bengal/ District/ Sub District, 1991 and 2001" (http://web.cmc.net.in/wbcensus/DataTables/02/Table4_18.htm). *West Bengal*. Directorate of census operations. . Retrieved 2008-12-03.

[9] "PARLIAMENTARY CONSTITUENCY MAP, South 24 Parganas" (http://www.s24pgs.gov.in/election/pc-map.html). . Retrieved 2011-04-17.

[10] "General election to the Legislative Assembly, 2001 – List of Parliamentary and Assembly Constituencies" (http://archive.eci.gov.in/ se2001/background/S25/WB_ACPC.pdf). *West Bengal*. Election Commission of India. . Retrieved 2008-11-21.

[11] "Press Note, Delimitation Commission" (http://www.wbgov.com/e-gov/English/DELIMITATION.pdf). *Assembly Constituencies in West Bengal*. Delimitation Commission. . Retrieved 2008-11-21.

[12] US Directorate of Intelligence. "Country Comparison:Population" (https://www.cia.gov/library/publications/the-world-factbook/ rankorder/2119rank.html). . Retrieved 2011-10-01. "Honduras 8,143,564"

[13] "2010 Resident Population Data" (http://2010.census.gov/2010census/data/apportionment-pop-text.php). U. S. Census Bureau. . Retrieved 2011-09-30. "Virginia 8,001,024"

[14] Indian Ministry of Forests and Environment. "Protected areas: West Bengal" (http://oldwww.wii.gov.in/envis/envis_pa_network/index. htm). . Retrieved September 25, 2011.

External links

- South 24 Parganas district official website - homepage (http://s24pgs.gov.in/index.htm)
- A travel article on Dhosa & Tilpi by Rangan Datta (http://www.rangan-datta.info/Dhosa &Tilpi.htm)
- A travel Article on Frazerganj by Rangan Datta (http://www.rangan-datta.info/Frazerganj.htm)
- A travel article on Achipur by Rangan Datta (http://www.rangan-datta.info/Achipur.htm)
- A travel article on Bowali by Rangan Datta (http://www.rangan-datta.info/bowali.html)
- Rangan Datta's Home Page (http://www.rangan-datta.info/index.htm)
- Pictures of Ganga Sagar (http://album.doorersathi.com/thumbnails.php?album=3)
- Pictures of Ganga Sagar Mela 2007 (http://album.doorersathi.com/thumbnails.php?album=19)
- Pictures of Ganga Sagar Mela 2008 (http://album.doorersathi.com/thumbnails.php?album=27)

Budge_Budge

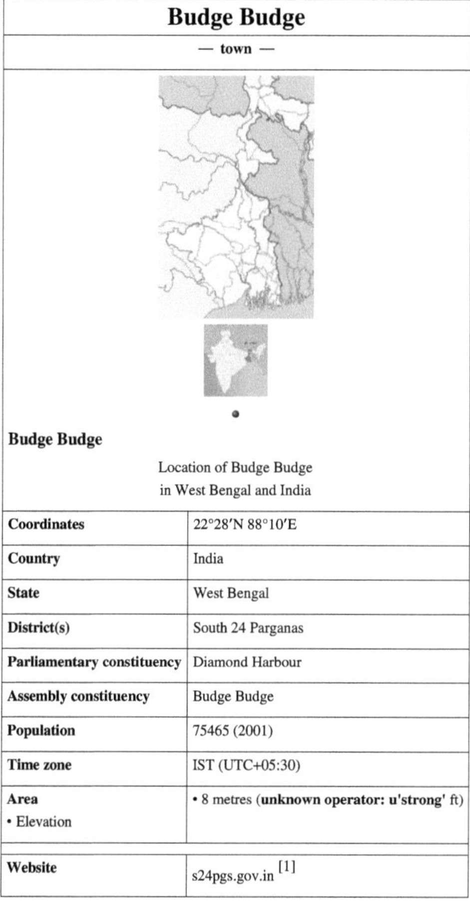

<table>
<tr><td colspan="2" align="center">Budge Budge</td></tr>
<tr><td colspan="2" align="center">— town —</td></tr>
<tr><td colspan="2" align="center">Budge Budge

Location of Budge Budge
in West Bengal and India</td></tr>
<tr><td>Coordinates</td><td>22°28′N 88°10′E</td></tr>
<tr><td>Country</td><td>India</td></tr>
<tr><td>State</td><td>West Bengal</td></tr>
<tr><td>District(s)</td><td>South 24 Parganas</td></tr>
<tr><td>Parliamentary constituency</td><td>Diamond Harbour</td></tr>
<tr><td>Assembly constituency</td><td>Budge Budge</td></tr>
<tr><td>Population</td><td>75465 (2001)</td></tr>
<tr><td>Time zone</td><td>IST (UTC+05:30)</td></tr>
<tr><td>Area
• Elevation</td><td>• 8 metres (unknown operator: u'strong' ft)</td></tr>
<tr><td>Website</td><td>s24pgs.gov.in [1]</td></tr>
</table>

Budge Budge (Bengali: বজ বজ *Bôj Bôj*) is a town and a municipality in South 24 Parganas district in the state of West Bengal, India. It is a part of the area covered by Kolkata Metropolitan Development Authority.[2]

History

Swami Vivekananda landed at Budge Budge ferry ghat when he returned from his Chicago visit. The anniversary is still celebrated today with great zeal.

A curious feature of this small and old town is the large number of Sikhs who live here. Budge Budge was the site where the ship Komagata Maru was allowed to land following its return from Vancouver. The ship was chartered by a group of Sikhs to challenge the exclusion laws enacted by Canada to restrict Indian immigration.

Budge Budge municipality is more than a century old establishment.

Historically the oldest people of this place were the 'Haldars' who came here to guard a fort near the bank of the River Ganges. A British writer who had come with Clive around 1740-50 chronicled this event. Maniklal was the main person at the fort and his soldiers lost to Clive's troops. This may have been the start of the series of Indian defeats to the expanding British colonial forces in the 18th century.

Geography

Budge Budge is located in the south-western suburbs of Calcutta, on the eastern bank of the River Ganges. Over the past few years Budge Budge has developed considerably in terms of lifestyle and infrastructure. With the ongoing project of construction of 4-lane road the economic progress is expected to accelerate.

Economics

Budge Budge owes much of its importance to the oil storage and jute mills. Being close to Calcutta and on the shores of Hooghly river makes it a strategic location for oil storage and is the biggest oil storage for the metropolis Calcutta with big PSUs like Petroleum Wharves Budge Budge (PWBB)under Kolkata Port trust. BPCL, HPCL, IOC having large units there. Jute mills were the biggest employers in the area till they started falling sick. Prominent among them are New Central Jute Mill and Budge Budge Jute Mills. At their height before 1971 these jute mills used to employ thousands of workers (New Central Jute Mills has been said to have employed as many as twenty thousand people) but after the partition of India and the subsequent creation of Bangladesh, supply of raw materials for these jute mills decreased. This, along with failure of trade unions lead to the closing of most of these jute mills.

The Budge Budge thermal power plant set up by CESC in Achipur (named after a Chinese called Achhu saheb by the locals who had established a sugar cane unit there) is a major source of electricity for Kolkata and its suburbs.

Transport

The Sealdah-Budge Budge line was constructed in 1890.[3] It is part of the Kolkata Suburban Railway system.

Ferry service from Bouria (Howrah District) to Budge Budge.

Demographics

As of 2001 India census,[4] Budge Budge had a population of 75,465. Males constitute 55% of the population and females 45%. Budge Budge has an average literacy rate of 70%, higher than the national average of 59.5%; with male literacy of 75% and female literacy of 64%. 10% of the population is under 6 years of age.

The majority of the population comprises Bengali Hindus belonging to Mahishya caste. Sunni Muslims live in specific areas. Christian families live in Hindu majority localities.

Education

Apart from the century old schools (Budge Budge Uchcha Balika Vidyalaya,P.K. High School,Kalipur High School, Sarengabad High School,Jagweshwari Paathshaala,Subhas Girl High School) several English medium schools (including the St Paul's day school, St Thomas, St. Stephenson, Carmel) have sprouted up but the quality of education is not of adequate standard and consequently, aspiring students are forced to look torwards Kolkata for better opportunities. The oldest school of this locality is Sarengabad High School. The school was established in 1856, the same year when Calcutta University was founded. Noted philanthropist of the locality late Shri Anath Bandhu Mitra was instrumental for converting the small village school to a large institution. Shri Mitra was also founder of Jagweshwari Paathshaala High School, Arya Rishikul School, Bandhab Pathagar and Baikuntha Ashram. His able grandson Dr.Ramesh Chandra Basu carried the legacy and during his tenure as the secretary, Sarangabad High School had a through transformation to its present form. There is a college named Budge Budge College [5].

Health care

High levels of arsenic in ground water was found in 12 blocks of the district. Water samples collected from tubewells in the affected places contained arsenic above the normal level (10 microgram a litre as specified by the World Health Organisation). The affected blocks are Baruipur, Bhangar I, Bhangar II, Bishnupur I, Bishnupur II, Basanti, Budge Budge, Canning I, Canning II, Sonarpur, Mograhat II and Joynagar. [6]

Nomenclature

Budge Budge has a reduplicated place name, similar to many such places in Australia.

References

[1] http://s24pgs.gov.in

[2] "Base Map of Kolkata Metroploitan area" (http://www.cmdaonline.com/kma.html). Kolkata Metropolitan Development Authority. . Retrieved 2007-09-03.

[3] Chaudhuri, Sukanta, *The Railway Comes to Calcutta*, in *Calcutta, the Living City*, Vol. I, edited by Saumya Chatterjee,budge budge.

[4] "Census of India 2001: Data from the 2001 Census, including cities, villages and towns (Provisional)" (http://web.archive.org/web/ 20040616075334/http://www.censusindia.net/results/town.php?stad=A&state5=999). Census Commission of India. Archived from the original (http://www.censusindia.net/results/town.php?stad=A&state5=999) on 2004-06-16. . Retrieved 2008-11-01.

[5] http://www.budgebudgecollege.org

[6] "High arsenic levels in South" (http://web.archive.org/web/20070929091658/http://www.thestatesman.net/page.arcview. php?clid=23&id=183781&usrsess=1). *The Statesman*, 24 June 2007. Archived from the original (http://www.thestatesman.net/page. arcview.php?clid=23&id=183781&usrsess=1) on 2007-09-29. . Retrieved 2007-09-03.

Balarampur,_Budgebudge

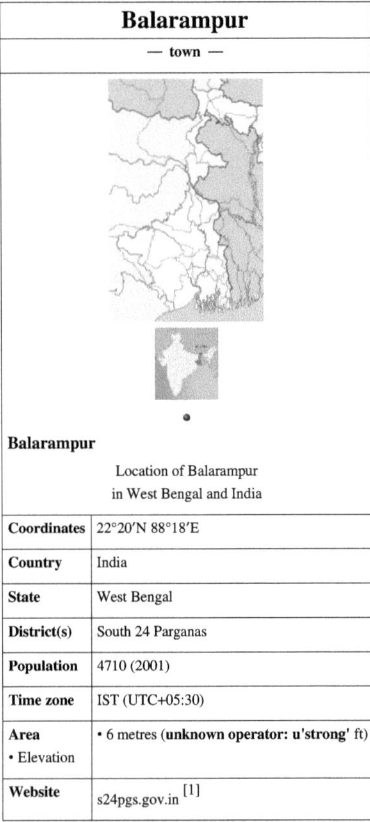

Balarampur	
— town —	
Balarampur	
Location of Balarampur in West Bengal and India	
Coordinates	22°20′N 88°18′E
Country	India
State	West Bengal
District(s)	South 24 Parganas
Population	4710 (2001)
Time zone	IST (UTC+05:30)
Area • Elevation	• 6 metres (**unknown operator: u'strong'** ft)
Website	s24pgs.gov.in [1]

Balarampur is a census town under Budgebudge police station [1] of Alipur sadar subdivision in South 24 Parganas district in the state of West Bengal, India.

Geography

Balarampur is located at 22°20′N 88°18′E[2] . It has an average elevation of 6 metres (20 feet).

Demographics

As of 2001 India census,[3] Balarampur had a population of 4710. Males constitute 51% of the population and females 49%. Balarampur has an average literacy rate of 64%, higher than the national average of 59.5%; with 56% of the males and 44% of females literate. 13% of the population is under 6 years of age.

References

[1] "Census of India" (http://web.archive.org/web/20070809031732/http://www.wbcensus.gov.in/DataTables/01/Table-3.htm). *District-wise list of stautory towns*. Directorate of census operations, West Bengal. Archived from the original (http://www.wbcensus.gov. in/DataTables/01/Table-3.htm) on 2007-08-09. . Retrieved 2007-09-02.

[2] Falling Rain Genomics, Inc - Balarampur (http://www.fallingrain.com/world/IN/28/Balarampur3.html)

[3] "Census of India 2001: Data from the 2001 Census, including cities, villages and towns (Provisional)" (http://web.archive.org/web/ 20040616075334/http://www.censusindia.net/results/town.php?stad=A&state5=999). Census Commission of India. Archived from the original (http://www.censusindia.net/results/town.php?stad=A&state5=999) on 2004-06-16. . Retrieved 2008-11-01.

Community_development_block_in_India

The **Community development block** (Hindi: सामुद्रायिक विकास ख·ड) is a rural area earmarked for administration and development in India. The area is administered by a Block Development Officer. A community development block covers several gram panchayats, local administrative unit at the village level.

History

The community development programme was launched on a pilot basis in 1952 to provide for a substantial increase in the country's agricultural programme, and for improvements in systems of communication, in rural health and hygiene, and in rural education. The community development programme was rapidly implemented. In 1956, by the end of the first five year plan period, there were 248 blocks, covering around a fifth of the population in the country. By the end the second five year plan period, there were 3,000 blocks covering 70 per cent of the rural population. By 1964, the entire country was covered.[1]

References

[1] "The Failure of the Community Development Programme in India" (http://cdj.oxfordjournals.org/cgi/pdf_extract/11/2/95). . Retrieved 2010-04-06.

Gram_panchayat

Gram panchayats are local self-governments at the village or small town level in India. As of 2002 there were about 265,000 gram panchayats in India. The gram panchayat is the foundation of the Panchayat System. A gram panchayat can be set up in villages with minimum population of 300. Sometimes two or more villages are clubbed together to form group-gram panchayat when the population of the individual villages is less than 300.

Indian municipal organisation by state

Structure

The Sarpanch or Chairperson is the head of the Gram Panchayat. The elected members of the Gram Panchayat elect from among themselves a Sarpanch and a Deputy Sarpanch for a term of five years. In some places the panchayat president is directly elected by village people. The Sarpanch presides over the meetings of the Gram Panchayat and supervises its working. He implements the development schemes of the village. The Deputy Sarpanch, who has the power to make his own decisions, assists the Sarpanch in his work.

The Sarpanch has the responsibilities of

1. Looking after street lights, construction and repair work of the roads in the villages and also the village markets, fairs,collection of tax, festivals and celebrations.
2. Keeping a record of births, deaths and marriages in the village.
3. Looking after public health and hygiene by providing facilities for sanitation and drinking water.
4. Providing for education.
5. To organize the meetings of Gramsabha(ग्रामसभा) and Grampanchayat(ग्रामपंचायत).

A grampannchyat consists of a between 7 and 17 members, elected from the wards of the village, and they are called a "panch". Peoples of village selects panch, with one-eighth of seats reserved for female candidates. To establish a Grampanchyat in a village, the population of the village should be at least 500 people of voting age.

Sources of income

The main source of income of the Gram Panchayat is the property tax levied on the buildings and the open spaces within the village. Other sources of income include professional tax, taxes on pilgrimage, animal trade, grant received from the State Government in proportion of land revenue and the grants received from the Zilla Parishad.

The gramsevak / gram vikas adhikari is communicator in government and village panchayat and do works for sarpanch. This post is in maharashtra, india. he is responsible person as sarpanch.

Principles of decentralisation

Dr S B Sen committee, a committee appointed by the Government of Kerala in 1996, had suggested the following principles, which was later adopted by the Second Administrative Reforms Commission, for local governance :-

- subsidiarity
- democratic decentralisation
- delineation of functions
- devolution of functions in real terms
- convergence
- citizen centricity

gram sabha is conducted six times in a year ...april, 1st may,15TH augest, 2octo, nov, 26 jan.

References

External links

- *Our Civic Life (Civics and Administration)*. Maharashtra State Bureau of Textbook Production and Curriculum Research, Pune
- Subramaniam Vincent (2002-02-28). "Ugly duckling to swan" (http://www.indiatogether.org/2003/apr/gov-kardcrefm.htm). India Together.

Scheduled_castes_and_scheduled_tribes

The **Scheduled Castes** (SCs), also known as the **Dalit**, and the **Scheduled Tribes** (STs) are two groupings of historically disadvantaged people that are given express recognition in the Constitution of India. During the period of British rule in the Indian sub-continent they were known as the **Depressed Classes**.

The Scheduled Castes and Scheduled Tribes make up around 15% and 7.5% respectively of the population of India, or around 24% altogether, according to the 2001 Census.[1] The proportion of Scheduled Castes and Scheduled Tribes in the country's population has steadily risen since independence in 1947.

The *Constitution (Scheduled Castes) Order, 1950* lists 1,108 castes across 25 states in its First Schedule,[2] while the *Constitution (Scheduled Tribes) Order, 1950* lists 744 tribes across 22 states in its First Schedule.[3]

Since Independence, the Scheduled Castes have benefited by the "Reservation" policy. This policy became an integral part of the Constitution through the efforts of Dr. Bhimrao Ambedkar, regarded as the father of the Indian constitution, who participated in Round Table Conferences and fought for the rights of the Depressed Classes. The Constitution lays down general principles for the policy of affirmative action for the SCs and STs.

History

From the 1850s these communities were loosely referred to as the "Depressed Classes". The early part of the 20th century saw a flurry of activity in the British Raj to assess the feasibility of responsible self-government for India. The Morley-Minto Reforms Report, Montagu–Chelmsford Reforms Report, and the Simon Commission were some of the initiatives that happened in this context. One of the hotly contested issues in the proposed reforms was the reservation of seats for the "depressed" classes in provincial and central legislatures.

In 1935 the British passed the Government of India Act 1935, designed to give Indian provinces greater self-rule and set up a national federal structure. Reservation of seats for the Depressed Classes was incorporated into the act, which came into force in 1937. The Act brought the term "Scheduled Castes" into use, and defined the group as including "such castes, races or tribes or parts of groups within castes, races or tribes, which appear to His Majesty in Council to correspond to the classes of persons formerly known as the 'Depressed Classes', as His Majesty in Council may prefer". This discretionary definition was clarified in *The Government of India (Scheduled Castes) Order, 1936,* which contained a list, or Schedule, of castes throughout the British administered provinces.

After independence, the Constituent Assembly continued the prevailing definition of Scheduled Castes and Tribes, and gave (via articles 341, 342) the President of India and Governors of states responsibility to compile a full listing of castes and tribes, and also the power to edit it later as required. The actual complete listing of castes and tribes was made via two orders *The Constitution (Scheduled Castes) Order, 1950,*[4] and *The Constitution (Scheduled Tribes) Order, 1950*[5] respectively.

Constitutional framework for safeguarding of interests

The Constitution provides a framework with a three pronged strategy [6] to improve the situation of SCs and STs.

1. Protective Arrangements - Such measures as are required to enforce equality, to provide punitive measures for transgressions, to eliminate established practices that perpetuate inequities, etc. A number of laws were enacted to operationalize the provisions in the Constitution. Examples of such laws include The Untouchability Practices Act, 1955, Scheduled Caste and Scheduled Tribe (Prevention of Atrocities) Act, 1989, The Employment of Manual scavengers and Construction of Dry Latrines (Prohibition) Act, 1993, etc.
2. Affirmative action - provide positive preferential treatment in allotment of jobs and access to higher education, as a means to accelerate the integration of the SCs and STs with mainstream society. Affirmative action is also popularly referred to as Reservation.
3. Development - Provide for resources and benefits to bridge the wide gap in social and economic condition between the SCs/STs and other communities.

National commissions

To effectively implement the various safeguards built into the Constitution and other legislations, the Constitution, under Articles 338 and 338A, provides for two statutory commissions - the **National Commission for Scheduled Castes**, and **National Commission for Scheduled Tribes**.

History

In the original Constitution, Article 338 provided for a Special Officer, called the Commissioner for SCs and STs, to have the responsibility of monitoring the effective implementation of various safeguards for SCs/STs in the Constitution as well as other related legislations and to report to the President. To enable efficient discharge of duties, 17 regional offices of the Commissioner were set up all over the country.

In the meanwhile there was persistent representation for a replacement of the Commissioner with a multi-member committee. It was proposed that the 48th Amendment to the Constitution be made to alter Article 338 to enable said proposal. While the amendment was being debated, the Ministry of Welfare issued an administrative decision to establish the Commission for SCs/STs as a multi-member committee to discharge the same functions as that of the Commissioner of SCs/STs. The first commission came into being in August 1978. The functions of the commission were modified in September 1987 to advise Government on broad policy issues and levels of development of SCs/STs.

In 1990 that the Article 338 was amended to give birth to the statutory National Commission for SCs and STs via the *Constitution (Sixty fifth Amendment) Bill, 1990.*[7] The first Commission under the 65th Amendment was constituted in March 1992 replacing the Commissioner for Scheduled Castes and Scheduled Tribes and the Commission set up under the Ministry of Welfare's Resolution of 1989.

In 2002, the Constitution was again amended to split the National Commission for Scheduled Castes and Scheduled Tribes into two separate commissions - the National Commission for Scheduled Castes and the National Commission for Scheduled Tribes.

Distribution

According to the 61st Round Survey of the NSSO, almost nine-tenths of Buddhists in India belonged to scheduled castes of the Constitution while one-third of Christians belonged to scheduled tribes. Major part of scheduled castes were Hindus by religion but belonged to castes and tribes having low population. The Sachar Committee report of 2006 also confirmed that members of scheduled castes and tribes of India are not exclusively adherents of Hinduism.

Religion	Scheduled Caste	Scheduled Tribe
Buddhism	89.50%	7.40%
Christianity	9.00%	32.80%
Sikhism	17.0%	0.90%
Hinduism	22.20%	9.10%
Gond	-	15.90%
Jainism	-	2.60%
Islam	0.80%	0.50%

Scheduled Caste Sub-Plan

The strategy of Scheduled Castes Sub-Plan (SCSP) of 1979 is an important intervention that mandates a planning process for social, economic, and educational development of Scheduled Castes and for improvement in their working and living conditions. It is an umbrella strategy that ensures flow of targeted financial and physical benefits from general sectors of development for the benefit of Scheduled Castes. Under this strategy, population.[8] It entails targeted flow of funds and associated benefits from the annual plan of States / Union Territories (UTs) at least in proportion to the SC population i.e. 16 % in the total population of the country / the particular state. Presently, 27 States / UTs having sizeable SC populations are implementing Scheduled Castes Sub-Plan. Although the Scheduled Castes population, according to 2001 Census, was 16.66 crores constituting 16.23% of the total population of India, the allocations made through SCSP in recent years have been much lower than the population proportion. Table hereafter provides the details of total State Plan Outlay, flow to Scheduled Castes Sub-Plan (SCSP) as reported by the State / UT Governments for the last few years especially since the present UPA government is in power at the

2004–2005	108788.9	17656	2065.38	11.06	68.3	5591
2005–2006	136234.5	22111	16422.63	12.05	74.3	5688
2006–2007	152088	24684	21461.12	14.11	86.9	3223
2007-2008*	155013.2	25159	22939.99	14.80	91.2	2219

- Information in respect of 14 States/UTs only and as on 31-12- 2007

Source: Network for Social Accountability (NSA) http://nsa.org.in

Prominent Personalities of SC/STs Community

- Guru Ravidas,North Indian Sant mystic of the bhakti movement
- Khusro Khan, or Khusru or Khusraw Khan was a medieval Indian military leader, and ruler of Delhi, as Sultan Nasir-ud-din, for a short period of time.He was a Dalit (Parwari-Mahar) caste from Gujrat. He converted to Islam from Hinduism at the time of his capture.[9] He was a untouchable in his own religion, but became a first Hindu to sit on the throne of Delhi.
- K. R. Narayanan, tenth President of India
- * Babu Jagjivan Ram, former Deputy Prime Minister of India
- B. R. Ambedkar, jurist, political leader, writer, father of Indian Constitution
- K. G. Balakrishnan, former Chief Justice of India
- Sushilkumar Shinde, Current Cabinet Minister for Power
- Prof. Nibaran Chandra Laskar, MP, Indian Parliament, was a dalit leader in Bengal and Assam.
- Mayavati, Chief Minister of Uttar Pradesh.
- Birsa Munda, Indian independence advocate, tribal leader and folk hero
- Damodaram Sanjivayya (1921–1972) (First dalit Chief Minister of a state in India and first dalit President of Indian National Congress party)
- Kanshi Ram, founder of Bahujan Samaj Party
- Dr. Mahendra Chandra Patni, a dalit leader and prominent scientist who has got the gold medal in LMP (the then MBBS) in 1923-24 batch from Berry-White School of Medicine, Dibrugarh, Assam, British India.
- D.Raja, Member of Rajyasabha,National Secratory for Communist Party of India
- G. M. C. Balayogi, dalit speaker, Lok Sabha,
- Ajit Jogi, first chief minister of the state of Chhattisgarh, India
- Shibu Soren, Ex Chief Minister of Jharkhand state in India
- Meira Kumar, Indian politician and Member of Parliament, Speaker of Lok Sabha
- S. Ashok Kumar, Judge Madras High Court and High Court of Andhra Pradesh
- Ram Vilas Paswan, the president of the Lok Janshakti Party, political party
- Bangaru Laxman, former President of Bharatiya Janata Party (BJP)
- Lala Ram Ken, Member of Parliament (7th and 8th), India
- Vinod Kambli, Indian cricketer
- Vinoo Mankad, Indian cricketer, He played in 44 Tests for India
- Thol. Thirumavalavan, Member of Parliament, The founder president of Viduthalai Chiruthaikal Katchi, Tamil Nadu
- Ilaiyaraaja, a noted music director and composer, Ilaiyaraaja is also a instrumentalist, conductor, and a songwriter
- E. Ponnuswamy, former M.O.S. Petroleum India.
- M.E.Loganathan, Municipal Commissioner, Government of Tamil Nadu.
- Damodar Raja Narasimha - Deputy Chief Minister of Andhra Pradesh
- Dr. J. Geeta Reddy - Leader of the House in the Legislative Assembly AP
- Amarjeet Bhagat MLA Sitapur
- Faguni Ram, Member of Parliament and Minister of State
- K. S. R. Murthy IAS, Former MP, Lok Sabha
- Prem Singh - MLA
- Late Lahori Ram Economic Development Commissioner California State and Founder member Guru Ravidass Sikh Gurdwara, Pittusburg
- Ram Lakha Former Mayor of Coventry
- Sardar Lakhbir Singh First Sikh Mayor Of Luton
- Giani Ditt Singh Ji Founder of Singh Sabha Movement

- Dr. Baldev Singh Sher First Dalit (Ravidasia/Ramdasia Sikh) Medical Graduate from Glasgow in 1910 and son of Giani Ditt Singh Ji
- Shaeed Baba Sangat Singh Ji Martyr in the Battle of Chamkaur Sahib
- Johnny Lever (Janumala John Prakasa Rao) - Famous Bollywood comedian, born in Vusullapalli near Kanigiri, Prakasm dt, AP.
- Betha Sudhakar ("Pichha kottudu sudhakar") - Popular Comedian & Character Artist in Tollywood
- Lankapalli Bullayya(1918–1992), former VC Andhra University(1968–74); first dalit to become the Vice-Chancellor of a university in India
- Late Shri Ram Ratan Ram— Member of Parliament (1984–1989)
- Dr.M.Velusamy (1973) is well known Social Science Scholar from Tamil Nadu. First Dalit Scholar Who has awarded his PhD at Madras Institute of Development Studies (MIDS), Chennai. His thesis entitled on Indian Constitution and Dalit Welfare : A Study of Tamil Nadu, 1950-2005. Published books on topic related to Dalits Periyar Dravidian Politics in Tamil Nadu.
- Jwala Prasad Kureel- MP of 6th Lok Sabha, Affiliated to Janata Party serving Ghatampur (UP) Lok Sabha Constituency
- Arun Anand- Former Scholar BTech Mech, MBA IIT Delhi, Social Worker, Running NGO for Dalit and poor in bangalore

See also

- List of Scheduled Tribes in India
- Scheduled Caste and Scheduled Tribe (Prevention of Atrocities) Act, 1989
- List of Sudra Hindu Saints
- Dalit saints of Hinduism
- Forward caste
- Other Backward Class
- Dalit Christian

Notes

[1] Census of India - India at a Glance : Scheduled Castes & Scheduled Tribes Population (http://www.censusindia.gov.in/ Census_Data_2001/India_at_Glance/scst.aspx)

[2] Text of the *Constitution (Scheduled Castes) Order, 1950*, as amended (http://lawmin.nic.in/ld/subord/rule3a.htm)

[3] Text of the *Constitution (Scheduled Tribes) Order, 1950*, as amended (http://lawmin.nic.in/ld/subord/rule9a.htm)

[4] THE CONSTITUTION (SCHEDULED CASTES) ORDER, 1950]1 (http://lawmin.nic.in/ld/subord/rule3a.htm)

[5] 1THE CONSTITUTION (SCHEDULED TRIBES) (http://lawmin.nic.in/ld/subord/rule9a.htm)

[6] http://nhrc.nic.in/Publications/reportKBSaxena.pdf

[7] "Constitution of India as of 29 July 2008" (http://lawmin.nic.in/coi/coiason29july08.pdf). *The Constitution Of India*. Ministry of Law & Justice. . Retrieved 13 April 2011.

[8] http://www.planningcommission.nic.in/plans/stateplan/scp&tsp/noteguidelinesFor.doc

External links

- Jobs for Tribals (http://tribaljobsindia.blogspot.com/)
- Dalit Indian Chamber of Commerce & Industry (http://www.dicci.org/index.html)
- Dalit and Adivasi Student Portal (http://www.scststudents.org)
- Rise of Dalit businessmen (http://ibnlive.in.com/news/dalit-becomes-entrepreneur-without-use-of-quotas/ 189929-3.html)
- Organization for SC & ST Govt Employees (http://www.ajjaks.com/)

West_Bengal

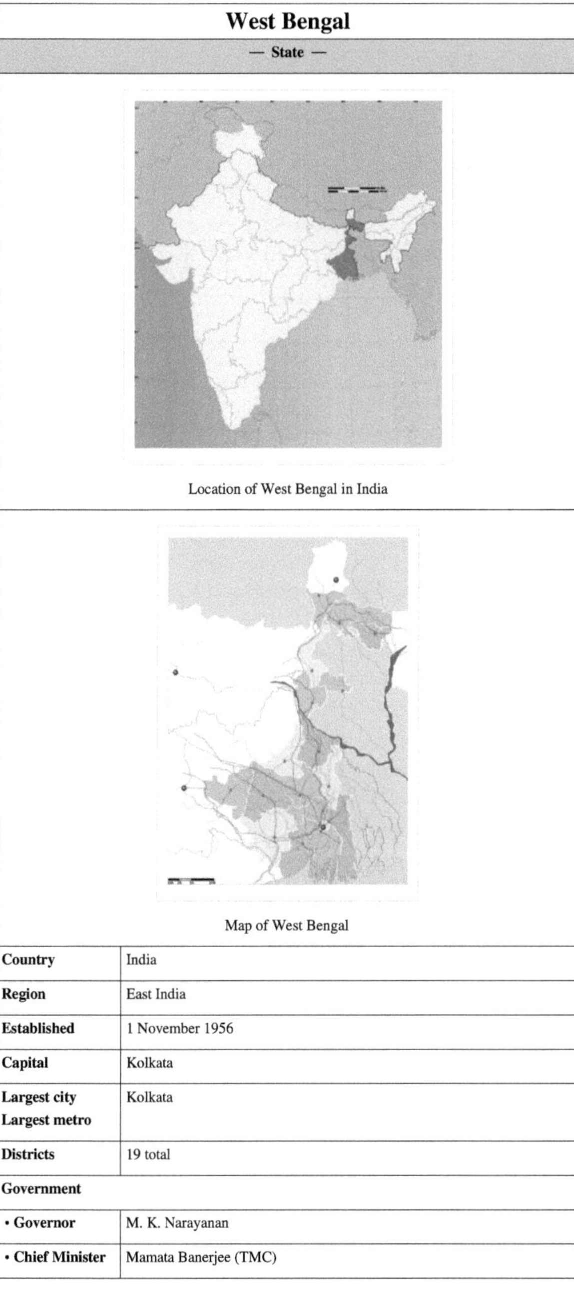

West Bengal	
— State —	

Location of West Bengal in India

Map of West Bengal

Country	India
Region	East India
Established	1 November 1956
Capital	Kolkata
Largest city Largest metro	Kolkata
Districts	19 total
Government	
• Governor	M. K. Narayanan
• Chief Minister	Mamata Banerjee (TMC)

• Legislature	Unicameral (295 • seats)
Area	
• Total	88752 km^2 (**unknown operator: u'strong'** sq mi)
Area rank	13th
Population (2011)[1]	
• Total	91347736
• Rank	4th
• Density	**unknown operator: u'strong'**/km^2 (**unknown operator: u'strong'**/sq mi)
Time zone	IST (UTC+05:30)
ISO 3166 code	IN-WB
HDI	▼ 0.625 (medium)
HDI rank	19th (2005)
Literacy	77.08%[2]
Official languages	Bengali · English
Website	westbengal.gov.in [3]
294 elected, 1 nominated	

West Bengal /bɛŋˈgɔːl/ (proposed new English name: *Paschim Banga*[4]) is a state in the eastern region of India and is the nation's fourth-most populous.[5] It is also the seventh-most populous sub-national entity in the world, with over 91 million inhabitants.[5] Covering a total area of 34267 sq mi (**unknown operator: u'strong'** km^2), it is bordered by the countries of Nepal, Bhutan, and Bangladesh, and the Indian states of Orissa, Jharkhand, Bihar, Sikkim, and Assam. The state capital is Kolkata (formerly *Calcutta*). West Bengal encompasses two broad natural regions: the Gangetic Plain in the south and the sub-Himalayan and Himalayan area in the north.

In the 3rd century BC, the broader region of Bengal was conquered by the emperor Ashoka. In the 4th century AD, it was absorbed into the Gupta Empire. From the 13th century onward, the region was ruled by several sultans, powerful Hindu states and Baro-Bhuyan landlords, until the beginning of British rule in the 18th century. The British East India Company cemented their hold on the region following the Battle of Plassey in 1757, and the city of Calcutta (now known as Kolkata) served for many years as the capital of British India. The early and prolonged exposure to British administration resulted in expansion of Western education, culminating in development in science, institutional education, and social reforms of the region, including what became known as the Bengal Renaissance. A hotbed of the Indian independence movement through the early 20th century, Bengal was divided in 1947 along religious lines into two separate entities: West Bengal – a state of India – and East Bengal, which initially joined the new nation of Pakistan, before becoming part of modern-day Bangladesh in 1971.

A major agricultural producer, West Bengal is the sixth-largest contributor to India's net domestic product.[6] Noted for its political activism, the state was ruled by democratically elected communist government for three decades. West Bengal is noted for its cultural activities, with the state capital Kolkata earning the sobriquet "cultural capital of India". The state's cultural heritage, besides folk culture, ranges from stalwarts in literature including Nobel-laureate Rabindranath Tagore to scores of musicians, film-makers and artist. West Bengal is also distinct from most other Indian states in its appreciation and practice of playing soccer besides the national favourite sport cricket.[7] [8] [9]

Etymology

The name of *Bengal*, or *Bangla*, is of unknown origins. Many theories have been formulated to explain the origin of the word "Bengal" or "Bangla". One theory suggests that the word derives from Dravidian tribes of 1000 B.C present at that time.[10] The word might have been derived from the ancient kingdom of *Vanga*, or *Banga*. Although some early Sanskrit literature mentions the name, the region's early history is obscure. The region was part of Mauryan empire under Ashoka in the 3rd century BCE.

History

Stone age tools dating back 20,000 years have been excavated in the state.[11] Remnants of civilization in the greater Bengal region date back four thousand years,[12] when the region was settled by Dravidian, Tibeto-Burman, and Austro-Asiatic peoples. The region was a part of the Vanga Kingdom, one of ancient kingdoms of Epic India. The kingdom of Magadha was formed in 7th century BC, consisting of the Bihar and Bengal regions. It was one of the four main kingdoms of India at the time of Mahavira and the Buddha, and consisted of several *Janapadas*.[13] During the rule of Maurya dynasty, the Magadha Empire extended over nearly all of South Asia, including Afghanistan and parts of Persia under Ashoka the Great in the 3rd century BC.

One of the earliest foreign references to Bengal is a mention of a land named Gangaridai by the Ancient Greeks around 100 BC. The word is speculated to have come from *Gangahrd* (Land with the Ganges in its heart) in reference to an area in Bengal.[14] Bengal had overseas trade relations with Suvarnabhumi (Burma, Lower Thailand, Lower Malay Peninsula, and the Sumatra).[15] According to Mahavamsa, Vijaya Singha, a Vanga prince, conquered Lanka (modern day Sri Lanka) and gave the name "Sinhala" to the country.[16]

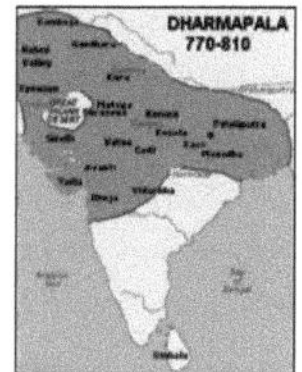

Pala Empire under Dharmapala   Pala Empire under Devapala

From the 3rd to the 6th centuries AD, the kingdom of Magadha served as the seat of the Gupta Empire. The first recorded independent king of Bengal was Shashanka, reigning around early 7th century.[17] After a period of anarchy, the Buddhist Pala dynasty ruled the region for four hundred years, followed by a shorter reign of the Hindu Sena dynasty. Islam made its first appearance in Bengal during the 12th century when Sufi missionaries arrived. Later, occasional Muslim raiders reinforced the process of conversion by building mosques, madrassas and Sufi Khanqah. Between 1202 and 1206, Bakhtiar Khilji, a military commander from the Delhi Sultanate, overran Bihar and Bengal as far east as Rangpur, Bogra and the Brahmaputra River. Although he failed to bring Bengal under his control, the expedition managed to defeat Lakshman Sen and his two sons moved to a place then called Vikramapur (present-day Munshiganj District), where their diminished dominion lasted until the late 13th century.

During the 14th century, the former kingdom became known as the Sultanate of Bengal, ruled intermittently with the Sultanate of Delhi as well as powerful Hindu states and land-lords-Baro-Bhuyans. The Hindu Deva Kingdom ruled over eastern Bengal after the collapse Sena Empire. The Sultanate of Bengal was interrupted by an uprising by the Hindus under Raja Ganesha. The Ganesha dynasty began in 1414, but his successors converted to Islam. Bengal came once more under the control of Delhi as the Mughals conquered it in 1576. There were several independent Hindu states established in Bengal during the Mughal period like those of Maharaja Pratap Aditya of Jessore and Raja Sitaram Ray of Burdwan. These kingdoms contributed greatly to the economic and cultural landscape of Bengal. Extensive land reclamations in forested and marshy areas were carried out and trade as well as commerce were highly encouraged. These kingdoms also helped introduce new music, painting, dancing and sculpture into Bengali art-forms as well as many temples were constructed during this period. Militarily, they served as bulwarks against Portuguese and Burmese attacks. Koch Bihar Kingdom in the northern Bengal, flourished during the period of 16th and the 17th centuries as well as weathered the Mughals and survived till the advent of the British.

Raja Ram Mohan Roy is widely regarded as the "Father of the Bengal Renaissance".

European traders arrived late in the fifteenth century. Their influence grew until the British East India Company gained taxation rights in Bengal *subah*, or province, following the Battle of Plassey in 1757, when Siraj ud-Daulah, the last independent Nawab, was defeated by the British.[18] The Bengal Presidency was established by 1765, eventually including all British territories north of the Central Provinces (now Madhya Pradesh), from the mouths of the Ganges and the Brahmaputra to the Himalayas and the Punjab. The Bengal famine of 1770 claimed millions of lives.[19] Calcutta was named the capital of British India in 1772. The Bengal Renaissance and Brahmo Samaj socio-cultural reform movements had great impact on the cultural and economic life of Bengal. The failed Indian rebellion of 1857 started near Calcutta and resulted in transfer of authority to the British Crown, administered by the Viceroy of India.[20] Between 1905 and 1911, an abortive attempt was made to divide the province of Bengal into two zones.[21] Bengal suffered from the Great Bengal famine in 1943 that claimed 3 million lives.[22]

Bengal played a major role in the Indian independence movement, in which revolutionary groups such as *Anushilan Samiti* and *Jugantar* were dominant. Armed attempts against the British Raj from Bengal reached a climax when Subhash Chandra Bose led the Indian National Army from Southeast Asia against the British. When India gained independence in 1947, Bengal was partitioned along religious lines. The western part went to India (and was named West Bengal) while the eastern part joined Pakistan as a province called East Bengal (later renamed East Pakistan, giving rise to independent Bangladesh in 1971).[23] Both West and East Bengal suffered from large refugee influx during the partition in 1947, leading to the political unrests later on. The partition of Bengal entailed the greatest exodus of people in Human History. Millions of Hindus migrated from East Pakistan to India and thousands of Muslims too went across the borders to East Pakistan. Because of the immigration of the refugees, there occurred the crisis of land and food in West Bengal; and such condition remained in long duration for more than three decades. The politics of West Bengal since the partition in 1947 developed round the nucleus of refugee

Subhash Chandra Bose was one of the most prominent Bengali freedom fighters in India's struggle for independence against the British Raj.

problem. Both the Rightists and the Leftists in the Politics of West Bengal have not yet become free from the socio-economic conditions created by the partition of Bengal. These conditions as have remained unresolved in some twisted forms have given birth to local socio-economic, political and ethnic movements.[24]

In 1950, the Princely State of Cooch Behar merged with West Bengal after King Jagaddipendra Narayan signed the Instrument of Accession with India.[25] In 1955, the former French enclave of Chandannagar, which had passed into Indian control after 1950, was integrated into West Bengal; portions of Bihar were subsequently merged with West Bengal.

During the 1970s and 1980s, severe power shortages, strikes and a violent Marxist-Naxalite movement damaged much of the state's infrastructure, leading to a period of economic stagnation. The Bangladesh Liberation War of 1971 resulted in the influx of millions of refugees to West Bengal, causing significant strains on its infrastructure.[26] The 1974 smallpox epidemic killed thousands. West Bengal politics underwent a major change when the Left Front won the 1977 assembly election, defeating the incumbent Indian National Congress. The Left Front, led by Communist Party of India (Marxist), governed for the state for the subsequent three decades.[27]

The state's economic recovery gathered momentum after economic reforms were introduced in the mid-1990s by the central government, aided by the advent of information technology and IT-enabled services. As of 2007, armed activists have been conducting minor terrorist attacks in some parts of the state,[28] [29] while clashes with the administration are taking place at several sensitive places over the issue of industrial land acquisition.[30] [31] Although the state's GDP has risen significantly since the 1990s, West Bengal has remained affected by political instability and bad governance.[32] The state continues to suffer from regular bandhs (strikes),[33] [34] a low Human Development Index level,[35] substandard healthcare services,[36] [37] a lack of socio-economic development,[38] poor infrastructure,[39] [40] political corruption and civil violence.[41] [42]

Geography and climate

West Bengal is on the eastern bottleneck of India, stretching from the Himalayas in the north to the Bay of Bengal in the south. The state has a total area of 88752 square kilometres (**unknown operator: u'strong'** sq mi).[1] The Darjeeling Himalayan hill region in the northern extreme of the state belongs to the eastern Himalaya. This region contains Sandakfu (3636 m or **unknown operator: u'strong'** ft)—the highest peak of the state.[43] The narrow Terai region separates this region from the plains, which in turn transitions into the Ganges delta towards the south. The Rarh region intervenes between the Ganges delta in the east and the western plateau and high lands. A small coastal region is on the extreme south, while the Sundarbans mangrove forests form a remarkable geographical landmark at the Ganges delta.

Many areas remain flooded during the heavy rains brought by monsoon

The Ganges is the main river, which divides in West Bengal. One branch enters Bangladesh as the *Padma* or *Pôdda*, while the other flows through West Bengal as the Bhagirathi River and Hooghly River. The Farakka barrage over Ganges feeds the Hooghly branch of the river by a feeder canal, and its water flow management has been a source of lingering dispute between India and Bangladesh.[44] The Teesta, Torsa, Jaldhaka and Mahananda rivers are in the northern hilly region. The western plateau region has rivers such as the Damodar,

Ajay and Kangsabati. The Ganges delta and the Sundarbans area have numerous rivers and creeks. Pollution of the Ganges from indiscriminate waste dumped into the river is a major problem.[45] Damodar, another tributary of the Ganges and once known as the "Sorrow of Bengal" (due to its frequent floods), has several dams under the Damodar Valley Project. At least nine districts in the state suffer from arsenic contamination of groundwater, and an estimated 8.7 million people drink water containing arsenic above the World Health Organisation recommended limit of 10 μg/L.[46]

National Highway 31A winds along the banks of the Teesta River near Kalimpong, in the Darjeeling Himalayan hill region.

West Bengal's climate varies from tropical savanna in the southern portions to humid subtropical in the north. The main seasons are summer, rainy season, a short autumn, and winter. While the summer in the delta region is noted for excessive humidity, the western highlands experience a dry summer like northern India, with the highest day temperature ranging from 38 °C (**unknown operator: u'strong' °F**) to 45 °C (**unknown operator: u'strong' °F**).[47] At nights, a cool southerly breeze carries moisture from the Bay of Bengal. In early summer brief squalls and thunderstorms known as *Kalbaisakhi*, or Nor'westers, often occur.[48] West Bengal receives the Bay of Bengal branch of the Indian ocean monsoon that moves in a northwest direction. Monsoons bring rain to the whole state from June to September. Heavy rainfall of above 250 cm is observed in the Darjeeling, Jalpaiguri and Cooch Behar district. During the arrival of the monsoons, low pressure in the Bay of Bengal region often leads to the occurrence of storms in the coastal areas. Winter (December–January) is mild over the plains with average minimum temperatures of 15 °C (**unknown operator: u'strong' °F**).[47] A cold and dry northern wind blows in the winter, substantially lowering the humidity level. However, the Darjeeling Himalayan Hill region experiences a harsh winter, with occasional snowfall at places.

Flora and fauna

State symbols of West Bengal

Union day	August 18 (Day of accession to India)	
State animal	Fishing cat[49]	
State bird	White-throated Kingfisher	
State tree	Devil Tree[49]	
State flower	Night-flowering Jasmine[49]	

A Bengal tiger.

Sal trees in the Arabari forest in West Midnapur.

As of 2009, recorded forest area in the state is 11879 km^2 (**unknown operator: u'strong'** sq mi) which is 13.38% of the state's geographical area,[50] compared to the national average of 21.02%.[51] [52] Reserves, protected and unclassed forests constitute 59.4%, 31.8% and 8.9%, respectively, of the forest area.[50] Part of the world's largest mangrove forest, the Sundarbans, is located in southern West Bengal.[53]

From a phytogeographic viewpoint, the southern part of West Bengal can be divided into two regions: the Gangetic plain and the littoral mangrove forests of the Sundarbans.[54] The alluvial soil of the Gangetic plain, compounded with favorable rainfall, make this region especially fertile.[54] Much of the vegetation of the western part of the state shares floristic similarities with the plants of the Chota Nagpur plateau in the adjoining state of Jharkhand.[54] The predominant commercial tree species is *Shorea robusta*, commonly known as the Sal tree. The coastal region of Purba Medinipur exhibits coastal vegetation; the predominant tree is the *Casuarina*. A notable tree from the Sundarbans is the ubiquitous *sundari* (*Heritiera fomes*), from which the forest gets its name.[55]

The distribution of vegetation in northern West Bengal is dictated by elevation and precipitation. For example, the foothills of the Himalayas, the *Dooars*, are densely wooded with Sal and other tropical evergreen trees.[56] However, above an elevation of 1000 metres (**unknown operator: u'strong'** ft), the forest becomes predominantly subtropical. In Darjeeling, which is above 1500 metres (**unknown operator: u'strong'** ft), temperate-forest trees such as oaks, conifers, and rhododendrons predominate.[56]

West Bengal has 3.26% of its geographical area under protected areas comprising 15 wildlife sanctuaries and 5 national parks[50] — Sundarbans National Park, Buxa Tiger Reserve, Gorumara National Park, Neora Valley National Park and Singalila National Park. Extant wildlife include Indian rhinoceros, Indian elephant, deer, bison, leopard, gaur, tiger, and crocodiles, as well as many bird species. Migratory birds come to the state during the winter.[57] The high-altitude forests of Singalila National Park shelter barking deer, red panda, chinkara, takin, serow, pangolin, minivet and Kalij pheasants. The Sundarbans are noted for a reserve project conserving the endangered Bengal tiger, although the forest hosts many other endangered species, such as the Gangetic dolphin, river terrapin and estuarine crocodile.[58] The mangrove forest also acts as a natural fish nursery, supporting coastal fishes along the Bay of Bengal.[58] Recognizing its special conservation value, Sundarban area has been declared as a Biosphere Reserve.[50]

Government and politics

West Bengal is governed through a parliamentary system of representative democracy, a feature the state shares with other Indian states. Universal suffrage is granted to residents. There are two branches of government. The legislature, the West Bengal Legislative Assembly, consists of elected members and special office bearers such as the Speaker and Deputy Speaker, that are elected by the members. Assembly meetings are presided over by the Speaker or the Deputy Speaker in the Speaker's absence. The judiciary is composed of the Calcutta High Court and a system of lower courts. Executive authority is vested in the Council of Ministers headed by the Chief Minister, although the titular head of government is the Governor. The Governor is the head of state appointed by the President of India. The leader of

Calcutta High Court is the highest court in West Bengal

the party or coalition with a majority in the Legislative Assembly is appointed as the Chief Minister by the Governor, and the Council of Ministers are appointed by the Governor on the advice of the Chief Minister. The Council of Ministers reports to the Legislative Assembly. The Assembly is unicameral with 295 Members of the Legislative Assembly, or MLAs,[59] including one nominated from the Anglo-Indian community. Terms of office run for 5 years, unless the Assembly is dissolved prior to the completion of the term. Auxiliary authorities known as *panchayats*, for which local body elections are regularly held, govern local affairs. The state contributes 42 seats to Lok Sabha[60] and 16 seats to Rajya Sabha of the Indian Parliament.[61]

The main players in the regional politics are the All India Trinamool Congress, the Indian National Congress, the Left Front alliance (led by the Communist Party of India (Marxist) or CPI(M)). Following the West Bengal State Assembly Election in 2011, the All India Trinamool Congress and Indian National Congress coalition under Mamata Banerjee of the All India Trinamool Congress was elected to power (getting 225 seats in the legislature).[62] West Bengal was ruled by the Left Front for the 34 years (1977–2011), making it the world's longest-running democratically elected communist government.[27]

Subdivisions

The 19 districts of West Bengal are as listed below.[63]

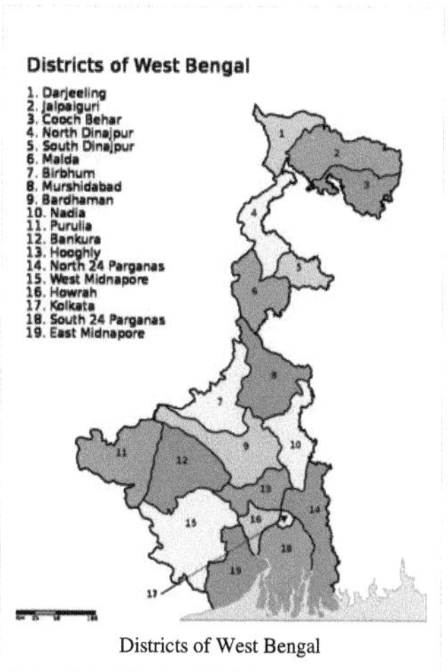

Districts of West Bengal

- Bankura
- Bardhaman
- Birbhum
- Cooch Behar
- Darjeeling
- East Midnapore
- Hooghly
- Howrah
- Jalpaiguri
- Kolkata
- Malda
- Murshidabad
- Nadia
- North 24 Parganas
- Uttar Dinajpur
- Purulia
- South 24 Parganas
- Dakshin Dinajpur
- West Midnapore

Each district is governed by a district collector or district magistrate, appointed either by the Indian Administrative Service or the West Bengal Civil Service.[64] Each district is subdivided into Sub-Divisions, governed by a sub-divisional magistrate, and again into Blocks. Blocks consists of panchayats (village councils) and town municipalities.[63]

The capital and largest city of the state is Kolkata – the third-largest urban agglomeration[65] and the seventh-largest city[66] in India. Asansol is the second largest city & urban agglomeration in West Bengal after Kolkata.[65] Siliguri is an economically important city, strategically located in the northeastern Siliguri Corridor (Chicken's Neck) of India. Other major cities and towns in West Bengal are Howrah, Durgapur, Raniganj, Haldia, Jalpaiguri, Kharagpur, Burdwan, Darjeeling, Midnapore, and Malda.[66]

Economy

Roadside vegetable vendor in a semi-rural area. A large proportion of residents are employed in informal sector.

Net State Domestic Product at Factor Cost at Current Prices (2004–05 Base)[6] figures in crores of Indian Rupees	
Year	Net State Domestic Product
2004–2005	190,073
2005–2006	209,642
2006–2007	238,625
2007–2008	272,166
2008–2009	309,799
2009–2010	366,318

In 2009–10, the tertiary sector of the economy (service industries) was the largest contributor to the gross domestic product of the state, contributing 57.8% of the state domestic product compared to 24% from primary sector (agriculture, forestry, mining) and 18.2% from secondary sector (industrial and manufacturing).[67] :12 Agriculture is the leading occupation in West Bengal. Rice is the state's principal food crop. Rice, potato, jute, sugarcane and wheat are the top five crops of the state.[67] :14 Tea is produced commercially in northern districts; the region is well known for Darjeeling and other high quality teas.[67] :14 State industries are localized in the Kolkata region, the mineral-rich western highlands, and Haldia port region.[68] The Durgapur–Asansol colliery belt is home to a number of major steel plants.[68] Manufacturing industries playing an important economic role are engineering products, electronics, electrical equipment, cables, steel, leather, textiles, jewellery, frigates, automobiles, railway coaches, and wagons. The Durgapur centre has established a number of industries in the areas of tea, sugar, chemicals and fertilizers. Natural resources like tea and jute in and nearby parts has made West Bengal a major centre for the jute and tea industries.

A significant part of the state is economically backward, namely, large parts of six northern districts of Cooch Behar, Darjeeling, Jalpaiguri, Malda, North Dinajpur and South Dinajpur; three western districts of Purulia, Bankura, Birbhum; and the Sundarbans area.[69] Years after independence, West Bengal was still dependent on the central government for meeting its demands for food; food production remained stagnant and the Indian green revolution bypassed the state. However, there has been a significant spurt in food production since the 1980s, and the state now has a surplus of grains.[69] The state's share of total industrial output in India was 9.8% in 1980–81, declining to 5% by 1997–98. However, the service sector has grown at a rate higher than the national rate.[69]

Freshly sown saplings of paddy; in the background are stacks of jute sticks

In terms net state domestic product (NSDP), West Bengal has the sixth largest economy (2009–2010) in India, with an NSDP of 366,318 crore Indian rupees, behind Maharashtra (817,891 crores), Uttar Pradesh (453,020 crores), Andhra Pradesh (426,816 crores), Tamil Nadu (417,716 crores), and Gujarat (370,400 crores).[6] In the period 2004–2005 to 2009–2010, the average gross state domestic product (GSDP) growth rate was 13.9% (calculated in Indian rupee term), lower than 15.5%, the average for all states of the country.[67] :4 The state's per capita GSDP at current prices in 2009–10 was US$ 956.4, improved from US$ 553.7 in 2004–05,[67] :10 but lower than the national per capita GSDP of US$ 1,302.[67] :4 The state's total financial debt stood at ₹ 191835 crore (US$38.27 billion) as of 2011.[70]

The state has promoted foreign direct investment, which has mostly come in the software and electronics fields; Kolkata is becoming a major hub for the Information technology (IT) industry. Rapid industrialisation process has given rise to debate over land acquisition for industry in this agrarian state.[71] NASSCOM–Gartner ranks West Bengal power infrastructure the best in the country.[72] Notably, many corporate companies are now headquartered in Kolkata include ITC Limited, India Government Mint, Kolkata, Haldia Petrochemicals, Exide Industries, Hindustan Motors, Britannia Industries, Bata India, Birla Corporation, CESC Limited, Coal India Limited, Damodar Valley Corporation, PwC India, Peerless Group, United Bank of India, UCO Bank and Allahabad Bank. In 2010s, events such as adoption of "Look East" policy by the government of India, opening of the Nathu La Pass in Sikkim as a border trade-route with China and immense interest in the South East Asian countries to enter the Indian market and invest have put Kolkata in an advantageous position for development in future, particularly with likes of Myanmar, where India needs oil from military regime.[73] [74]

Transport

Kolkata Suburban Railway caters to the commuters of the populous suburbs of Kolkata

As of 2011, the total length of surface road in West Bengal is over 92023 km (**unknown operator: u'strong'** mi);[67] :18 national highways comprise 2578 km (**unknown operator: u'strong'** mi)[75] and state highways 2393 km (**unknown operator: u'strong'** mi).[67] :18 As of 2006, the road density of the state is 103.69 km per 100 km² (166.92 mi per 100 sq mi), higher than the national average of 74.7 km per 100 km² (120 mi per 100 sq mi).[76] Average speed on state highways varies between 40–50 km/h (25–31 mi/h); in villages and towns, speeds are as low as 20–25 km/h (12–16 mi/h) due to the substandard quality of road constructions and low maintenance. As of 2011, the total railway route length is around 4481 km (**unknown operator: u'strong'** mi).[67] :20 Kolkata is the headquarters of two divisions of the Indian Railways—Eastern Railway and South Eastern Railway.[77] The Northeast Frontier Railway (NFR) plies in the northern parts of the state. The Kolkata metro is the country's first underground railway.[78] The Darjeeling Himalayan Railway, part of NFR, is a UNESCO World Heritage Site.[79]

The state's only international airport is Netaji Subhash Chandra Bose International Airport at Dum Dum, Kolkata. Bagdogra airport near Siliguri is another significant airport in the state. Kolkata is a major river-port in eastern India. The Kolkata Port Trust manages both the Kolkata docks and the Haldia docks.[80] There is passenger service to Port Blair on the Andaman and Nicobar Islands and cargo ship service to ports in India and abroad, operated by the Shipping Corporation of India. Ferry is a principal mode of transport in the southern part of the state, especially in the Sundarbans area. Kolkata is the only city in India to have trams as a mode of transport and these are operated by the Calcutta Tramways Company.[81]

Several government-owned organisations operate substandard bus services in the state, including the Calcutta State Transport Corporation, the North Bengal State Transport Corporation, the South Bengal State Transport Corporation, the West Bengal Surface Transport Corporation, and the Calcutta Tramways Company, thus leading to mismanagement. There are also private bus companies. The railway system is a nationalised service without any private investment. Hired forms of transport include metered taxis and auto rickshaws which often ply specific routes in cities. In most of the state, cycle rickshaws, and in Kolkata, hand-pulled rickshaws, are also used for short-distance travel. Large-scale transport accidents in West Bengal are common, particularly the sinking of transport boats and train crashes.[82]

Demographics

Dakshineswar Kali Temple

Tipu Sultan Mosque

According to the provisional results of 2011 national census, West Bengal is the fourth most populous state in India with a population of 91,347,736 (7.55% of India's population).[1] Majority of the population comprises Bengalis.[84] The Marwaris, Bihari and Oriya minority are scattered throughout the state; communities of Sherpas and ethnic Tibetans can be found in Darjeeling Himalayan hill region. Darjeeling district has a large number of Gurkha people of Nepalese origin. West Bengal is home to indigenous tribal *Adivasis* such as Santals, Kol, Koch-Rajbongshi and Toto tribe. There are a small number of ethnic minorities primarily in the state capital, including Chinese, Tamils, Gujaratis, Anglo-Indians, Armenians, Punjabis, and Parsis.[85] India's sole Chinatown is in eastern Kolkata.[86]

Religions in West Bengal[87]	
Religion	Percent
Hindu	72.5%
Muslim	25.2%
Others	2.3%

The official language is Bengali and English.[88] Nepali is the official language in three subdivisions of Darjeeling district.[88] As of 2001, in descending order of number of speakers, the languages of the state are: Bengali, Hindi, Santali, Urdu, Nepali, and Oriya.[88] Languages such as Rajbongshi and Ho are used in some parts of the state.

As of 2001, Hinduism is the principal religion at 72.5% of the total population, while Muslims comprise 25.2% of the total population , being the second-largest community as also the largest minority group; Sikhism, Christianity and other religions make up the remainder.[87] The state contributes 7.8% of India's population.[89] The state's 2001–2011 decennial growth rate was 13.93%,[1] lower than 1991–2001 growth rate of 17.8%,[1] and also lower than the national rate of 17.64%.[90] The gender ratio is 947 females per 1000 males.[90] As of 2011, West Bengal has a population density of 1029 inhabitants per square kilometre (**unknown operator: u'strong'** /sq mi) making it the second-most densely populated state in India, after Bihar.[90]

The literacy rate is 77.08%, higher than the national rate of 74.04%.[91] Data of 1995–1999 showed the life expectancy in the state was 63.4 years, higher than the national value of 61.7 years.[92] About 72% of people live in rural areas. The proportion of people living below the poverty line in 1999–2000 was 31.9%.[69] Scheduled Castes and Tribes form 28.6% and 5.8% of the population respectively in rural areas, and 19.9% and 1.5% respectively in urban areas.[69]

The crime rate in the state in 2004 was 82.6 per 100,000, which was half of the national average.[93] This is the fourth-lowest crime rate among the 32 states and union territories of India.[94] However, the state reported the highest rate of Special and Local Laws (SLL) crimes.[95] In reported crimes against women, the state showed a crime rate of 7.1 compared to the national rate of 14.1.[94] Some estimates state that there are more than 60,000 brothel-based women and girls in prostitution in Kolkata.[96] The population of prostitutes in Sonagachi constitutes mainly of Nepalese, Indians and Bangladeshis.[96] Some sources estimate there are 60,000 women in the brothels of Kolkata.[96] The largest prostitution area in city is Sonagachi.[96] West Bengal was the first Indian state to constitute a Human Rights Commission of its own.[94]

Culture

Literature

Rabindranath Tagore is Asia's first Nobel laureate and composer of India's national anthem

Swami Vivekananda was a key figure in introducing Vedanta and Yoga in Europe and USA,[] raising interfaith awareness and making Hinduism a world religion.[]

The Bengali language boasts a rich literary heritage, shared with neighbouring Bangladesh. West Bengal has a long tradition in folk literature, evidenced by the *Charyapada*, *Mangalkavya*, *Shreekrishna Kirtana*, *Thakurmar Jhuli*, and stories related to Gopal Bhar. In the nineteenth and twentieth century, Bengali literature was modernized in the works of authors such as Bankim Chandra Chattopadhyay, Michael Madhusudan Dutt, Rabindranath Tagore, Kazi Nazrul Islam, Sharat Chandra Chattopadhyay, Jibananda Das and Manik Bandyopadhyay. In modern times Jibanananda Das, Bibhutibhushan Bandopadhyay, Tarashankar Bandopadhyay, Manik Bandopadhyay, Ashapurna Devi, Shirshendu Mukhopadhyay, Buddhadeb Guha, Mahashweta Devi, Samaresh Majumdar, Sanjeev Chattopadhyay and Sunil Gangopadhyay among others are well known.

Music and dance

Baul singers at Basanta-Utsab, Shantiniketan

The Baul tradition is a unique heritage of Bengali folk music, which has also been influenced by regional music traditions.[97] Other folk music forms include Gombhira and Bhawaiya. Folk music in West Bengal is often accompanied by the ektara, a one-stringed instrument. West Bengal also has an heritage in North Indian classical music. "Rabindrasangeet", songs composed and set into tune by Rabindranath Tagore and "Nazrul geeti" (by Kazi Nazrul Islam) are popular. Also prominent are other musical forms like Dwijendralal, Atulprasad and Rajanikanta's songs, and *"adhunik"* or modern music from films and other composers.

From the early 1990s, there has been an emergence and popularisation of new genres of music, including fusions of Baul and Jazz by several Bangla bands, as well as the emergence of what has been called *Jeebonmukhi Gaan* (a modern genre based on realism). Bengali dance forms draw from folk traditions, especially those of the tribal groups, as well as the broader Indian dance traditions. Chau dance of Purulia is a rare form of mask dance. State is known for Bengali folk music such as baul and kirtans and *gajan*, and modern songs including Bengali adhunik songs. From the early 1990s, there has been an emergence of new genres of music, including the emergence of what has been called

Dance with Rabindra Sangeet.

Bengali *Jeebonmukhi Gaan* (a modern genre based on realism) by artists like Anjan Dutta, Kabir Suman, Nachiketa and folk/alternative/rock bands like Moheener Ghoraguli, Chandrabindoo, Bhoomi, Cactus and Fossils. Dutta's songs are influenced by classical music, and especially country music and blues and Bob Dylan and Leonard Cohen which he fused with Bengali tradition of east west, as did Suman. American urban folk and grunge are also an inspiration for this generation.[98]

Films

Mainstream Hindi films are popular in Bengal, and the state is home to a thriving cinema industry, dubbed "Tollywood". Tollygunj in Kolkata is the location of numerous Bengali movie studios, and the name "Tollywood" (similar to Hollywood and Bollywood) is derived from that name. The Bengali film industry is well known for its art films, and has produced acclaimed directors like Satyajit Ray, Mrinal Sen, Tapan Sinha and Ritwik Ghatak. Prominent contemporary directors include Buddhadev Dasgupta, Goutam Ghose, Aparna Sen and Rituparno Ghosh.

Fine arts

Bengal had been the harbinger of modernism in fine arts. Abanindranath Tagore, called the father of Modern Indian Art had started the Bengal School of Art which was to create styles of art outside the European realist tradition which was taught in art colleges under the colonial administration of the British Government. The movement had many adherents like Gaganendranath Tagore, Ramkinkar Baij, Jamini Roy and Rabindranath Tagore. After Indian Independence, important groups like the Calcutta Group and the Society of Contemporary Artists were formed in Bengal which dominated the art scene in India.

Reformist heritage

The capital, Kolkata, was the workplace of several social reformers, like Raja Ram Mohan Ray, Iswar Chandra Vidyasagar, and Swami Vivekananda. These social reforms have eventually led to a cultural atmosphere where practices like sati, dowry, and caste-based discrimination or untouchability, the evils that crept into the Hindu society, were abolished.

Cuisine

Rice and fish are traditional favourite foods, leading to a saying in Bengali, *machhe bhate bangali*, that translates as "fish and rice make a Bengali".[99] Bengal's vast repertoire of fish-based dishes includes hilsa preparations, a favorite among Bengalis. There are numerous ways of cooking fish depending on the texture, size, fat content and the bones. Sweets occupy an important place in the diet of Bengalis and at their social ceremonies. It is an ancient custom among both Hindu and Muslim Bengalis to distribute sweets during festivities. The confectionery industry has flourished because of its close association with social and religious

Patisapta - A kind of Pitha; which is a popular sweet dish in West Bengal during winter.

ceremonies. Competition and changing tastes have helped to create many new sweets. Bengalis make distinctive sweetmeats from milk products, including *Rôshogolla, Chômchôm, Kalojam* and several kinds of *sondesh*. Pitha, a kind of sweet cake, bread or dimsum are specialties of winter season. Sweets like coconut-naru, til-naru, moa, payesh, etc. are prepared during the festival of Lakshmi puja. Popular street food includes *Aloor Chop*, Beguni, Kati roll, and phuchka.[100] [101]

The variety of fruits and vegetables that Bengal has to offer is incredible. A host of gourds, roots and tubers, leafy greens, succulent stalks, lemons and limes, green and purple eggplants, red onions, plantain, broad beans, okra, banana tree stems and flowers, green jackfruit and red pumpkins are to be found in the markets or anaj bazaar as popularly called. *Panta bhat* (rice soaked overnight in water)with onion & green chili is a traditional dish consumed in rural areas. Common spices found in a Bengali kitchen are cumin, ajmoda (radhuni), bay leaf, mustard, ginger, green chillies, turmeric, etc. People of erstwhile East Bengal use a lot of ajmoda, coriander leaves, tamarind, coconut and mustard in their cooking; while those aboriginally from West Bengal use a lot of sugar, garam masala and red chilli powder. Vegetarian dishes are mostly without onion and garlic.

A *Murti* (representation) of Maa Durga

Costumes

Bengali women commonly wear the *shaṛi* , often distinctly designed according to local cultural customs. In urban areas, many women and men wear Western attire. Among men, western dressing has greater acceptance. Men also wear traditional costumes such as the *panjabi* with *dhuti*, often on cultural occasions.

Festivals

Durga Puja in October is the most popular festival in the West Bengal.[102] Poila Baishakh (the Bengali New Year), Rathayatra, Dolyatra or Basanta-Utsab, Nobanno, *Poush Parbon* (festival of Poush), Kali Puja, SaraswatiPuja, LaxmiPuja, Christmas, Eid ul-Fitr, Eid ul-Adha and Muharram are other major festivals. Buddha Purnima, which marks the birth of Gautama Buddha, is one of the most important Hindu/Buddhist festivals while Christmas, called *Bôṛodin* (Great day) in Bengali is celebrated by the minority Christian population. Poush mela is a popular festival of Shantiniketan, taking place in winter. West Bengal has been home to several famous religious teachers, including Sri Chaitanya, Sri Ramakrishna, Swami Vivekananda, A. C. Bhaktivedanta Swami Prabhupada and Paramahansa Yogananda. The *swami* is credited with introducing Hinduism to western society and became a religious symbol of the nation in the eyes of the intellectuals of the west.

Education

West Bengal schools are run by the state government or by private organisations, including religious institutions. Instruction is mainly in English or Bengali, though Urdu is also used, especially in Central Kolkata. The secondary schools are affiliated with the Council for the Indian School Certificate Examinations (CISCE), the Central Board for Secondary Education (CBSE), the National Institute of Open School (NIOS) or the West Bengal Board of Secondary Education.[103] Under the 10+2+3 plan, after completing secondary school, students typically enroll for 2 years in a junior college, also known as pre-university, or in schools with a higher secondary facility affiliated with the West Bengal Council of Higher Secondary Education or any central board. Students choose from one of three streams, namely liberal arts, commerce or science. Upon completing the required coursework, students may enroll in general or professional degree programs.

Medical College Kolkata

St. Paul's School, Darjeeling (around 1905)

IIT Kharagpur

West Bengal has eighteen universities.[104] [105] The University of Calcutta, one of the oldest and largest public universities in India, has more than 200 affiliated colleges. Kolkata has played a pioneering role in the development of the modern education system in India. It is the gateway to the revolution of European education. Sir William Jones (philologist) established the Asiatic Society in 1794 for promoting oriental studies. People like Ram Mohan Roy, David Hare, Ishwar Chandra Vidyasagar and William Carey played a leading role in the setting up of modern schools and colleges in the city. The Fort William College was established in 1810. The Hindu College was established in 1817. In 1855 the Hindu College was renamed as the Presidency

College.[106] The Bengal Engineering & Science University and Jadavpur University are prestigious technical universities.[107] Visva-Bharati University at Santiniketan is a central university and an institution of national importance.[108] The state has several higher education institutes of national importance including Indian Institute of Foreign Trade, Indian Institute of Management Calcutta (the first IIM), Indian Institute of Science Education and Research, Kolkata, Indian Statistical Institute, Indian Institute of Technology Kharagpur (the first IIT), National Institute of Technology, Durgapur and West Bengal National University of Juridical Sciences. After 2003 the state govt supported the creation of West Bengal University of Technology, West Bengal State University and Gour Banga University.

Besides these, the state also has Kalyani University, The University of Burdwan, Vidyasagar University and North Bengal University-all well established and nationally renowned, to cover the educational needs at the district levels of the state and also an Indian Institute of Science Education and Research, Kolkata. Also recently Presidency College, Kolkata became a University named Presidency University. Apart from this there is another private university run by Ramakrishna mission named Ramakrishna Mission Vivekananda University at Belur Math. There are a number of research institutes in kolkata. The Indian Association for the Cultivation of Science is the first research institute

IIM Calcutta

in Asia. C. V. Raman got Nobel Prize for his discovery (Raman Effect) done in IACS. Also Bose Institute, Saha Institute of Nuclear Physics, S.N. Bose National Centre for Basic Sciences, Indian Institute of Chemical Biology, Central Glass and Ceramic Research Institute, Variable Energy Cyclotron Center are most prominent. A large number of Indian Scholars are educated at different universities in Bengal. State has produced likes of Jagadish Chandra Bose, Satyendra Nath Bose and RC Bose.

Media

West Bengal had 505 published newspapers in 2005,[109] of which 389 were in Bengali.[109] *Ananda Bazar Patrika*, published from Kolkata with 1,277,801 daily copies, has the largest circulation for a single-edition, regional language newspaper in India.[109] Other major Bengali newspapers are *Bartaman, Sangbad Pratidin, Aajkaal, Jago Bangla, Uttarbanga Sambad* and *Ganashakti*. Major English language newspapers which are published and sold in large numbers are *The Telegraph, The Times of India, Hindustan Times, The Hindu, The Statesman, The Indian Express* and *Asian Age*. Some prominent financial dailies like *The Economic Times, Financial Express, Business Line* and *Business Standard* are widely circulated. Vernacular newspapers such as those in Hindi, Nepali Gujarati, Oriya, Urdu and Punjabi are also read by a select readership.

Doordarshan is the state-owned television broadcaster. Multi system operators provide a mix of Bengali, Nepali, Hindi, English and international channels via cable. Bengali 24-hour television news channels include STAR Ananda, Tara Newz, Kolkata TV, News Time, 24 Ghanta, Mahuaa Khobor, Ne Bangla, CTVN Plus, Channel 10 and R Plus.[110] [111] All India Radio is a public radio station.[111] Private FM stations are available only in cities like Kolkata, Siliguri and Asansol.[111] Vodafone, Airtel, BSNL, Reliance Communications, Uninor, Aircel, MTS India, Tata Indicom, Idea Cellular and Tata DoCoMo are available cellular phone operators. Broadband internet is available in select towns and cities and is provided by the state-run BSNL and by other private companies. Dial-up access is provided throughout the state by BSNL and other providers.

Sports

Salt Lake Stadium – Yuva Bharati Krirangan, Kolkata

Cricket and football (soccer) are popular sports in the state. West Bengal, unlike most other states of India, is noted for its passion and patronage of football.[7] [8] [9] Kolkata is one of the major centers for football in India[112] and houses top national clubs such as East Bengal, Mohun Bagan and Mohammedan Sporting Club.[113] Indian sports such as Kho Kho and Kabaddi are also played. Calcutta Polo Club is considered as the oldest polo club of the world,[114] and the Royal Calcutta Golf Club is the oldest of its kind outside Great Britain.[115]

West Bengal has several large stadiums—The Eden Gardens is one of only two 100,000-seat cricket amphitheaters in the world, although renovations will reduce this figure.[116] Kolkata Knight Riders, East Zone and Bengal play there, and the 1987 World Cup final was there although in 2011 World Cup, Eden Gardens was stripped due to construction incompleteness. Salt Lake Stadium—a multi-use stadium—is the world's second highest-capacity football stadium.[117] [118] Calcutta Cricket and Football Club is the second-oldest cricket club in the world.[119] National and international sports events are also held in Durgapur, Siliguri and Kharagpur.[120] Notable sports persons

Eden Gardens in Kolkata

from West Bengal include former Indian national cricket captain Sourav Ganguly, Pankaj Roy Olympic tennis bronze medallist Leander Paes, and chess grand master Dibyendu Barua. Other major sporting icons over the years include famous football players such as Chuni Goswami, PK Banerjee and Sailen Manna as well as swimmer Mihir Sen and athlete Jyotirmoyee Sikdar (winner of gold medals at the Asian Games).[121]

See also

- Bengal Army
- Bengali calendar
- Bengali Language Movement
- History of India
- List of people from West Bengal

Notes

[1] "Area, population, decennial growth rate and density for 2001 and 2011 at a glance for West Bengal and the districts: provisional population totals paper 1 of 2011: West Bengal" (http://www.censusindia.gov.in/2011-prov-results/prov_data_products_wb.html). Registrar General & Census Commissioner, India. . Retrieved 26 January 2012.

[2] "Sex ratio, 0-6 age population, literates and literacy rate by sex for 2001 and 2011 at a glance for West Bengal and the districts: provisional population totals paper 1 of 2011: West Bengal" (http://www.censusindia.gov.in/2011-prov-results/prov_data_products_wb.html). Government of India:Ministry of Home Affairs. . Retrieved 29 January 2012.

[3] http://www.westbengal.gov.in/

[4] Special correspondent (19 August 2011). "West Bengal may be renamed PaschimBanga" (http://www.thehindu.com/news/national/article2373155.ece). *The Hindu* (Chennai, India). . Retrieved 7 February 2012.

[5] "India: Administrative Divisions (population and area)" (http://www.world-gazetteer.com/wg.php?x=&men=gadm&lng=en&des=wg&geo=-104&srt=npan&col=abcdefghinoq&msz=1500&va=x). Census of India. . Retrieved April 17, 2009.

[6] "Net state domestic product at factor cost—state-wise (at current prices)" (http://www.rbi.org.in/scripts/PublicationsView.aspx?id=13592). *Handbook of statistics on Indian economy*. Reserve Bank of India. 15 September 2011. . Retrieved 7 February 2012.

[7] Dineo, Paul; Mills, James (2001). *Soccer in South Asia: empire, nation, diaspora*. London: Frank Cass Publishers. p. 71. ISBN 978-0-7146-8170-2.

[8] Bose, Mihir (2006). *The magic of Indian cricket: cricket and society in India*. Psychology Press. p. 240. ISBN 9780415356916.

[9] Das Sharma, Amitabha (2002). "Football and the big fight in Kolkata" (http://www.la84foundation.org/SportsLibrary/FootballStudies/2002/FS0502g.pdf) (PDF). *Football Studies* **5** (2): 57. . Retrieved `5 April 2012.

[10] "Bangldesh: early history, 1000 B.C.–A.D. 1202" (http://memory.loc.gov/cgi-bin/query/r?frd/cstdy:@field(DOCID+bd0014)). *Bangladesh: A country study*. Washington, D.C.: Library of Congress. September 1988. . Retrieved 2 March 2012. "Historians believe that Bengal, the area comprising present-day Bangladesh and the Indian state of West Bengal, was settled in about 1000 B.C. by Dravidian-speaking peoples who were later known as the Bang. Their homeland bore various titles that reflected earlier tribal names, such as Vanga, Banga, Bangala, Bangal, and Bengal."

[11] Sarkar, Sebanti (March 28, 2008). "History of Bengal just got a lot older" (http://www.telegraphindia.com/1080328/jsp/frontpage/story_9067406.jsp). *The Telegraph* (Calcutta, India). . Retrieved 13 September 2010. "Humans walked on Bengal's soil 20,000 years ago, archaeologists have found out, pushing the state's pre-history back by some 8,000 years."

[12] Bharadwaj, G (2003). "The Ancient Period". In Majumdar, RC. *History of Bengal*. B.R. Publishing Corp.

[13] Sultana, Sabiha. "Settlement in Bengal (Early Period)" (http://www.banglapedia.org/httpdocs/HT/S_0221.HTM). *Banglapedia*. Asiatic Society of Bangladesh. . Retrieved 4 March 2012.

[14] Chowdhury, AM. "Gangaridai" (http://banglapedia.search.com.bd/HT/G_0019.htm). *Banglapedia*. Asiatic Society of Bangladesh. . Retrieved 8 September 2006.

[15] Prasad, Prakash Chandra (2003). *Foreign trade and commerce in ancient India* (http://books.google.com/?id=mFW3sXnzEQ4C&pg=PA231&dq=ancient+history+of+bengal+trade#v=onepage&q=bengal&f=false). New Delhi: Abhinav Publications. p. 28. ISBN 978-81-7017-053-2. . Retrieved 13 September 2010.

[16] Geiger, Wilhelm (2003) [1908]. "Chapter VI: The Coming of Viajaya" (http://lakdiva.org/mahavamsa/chap006.html). *Mahavamsa: Great Chronicle of Ceylon* (http://books.google.com/?id=nX2af3kcregC&printsec=frontcover&dq=wilhelm+geiger#v=onepage&q&f=false). New Delhi: Asian Educational Services. pp. 51–54. ISBN 81-206-0218-8. . Retrieved 2 March 2012.

[17] Bhattacharyya, P.K.. "Shashanka" (http://www.banglapedia.org/httpdocs/HT/S_0122.HTM). *Banglapedia*. Asiatic Society of Bangladesh. . Retrieved 2 March 2012.

[18] Chaudhury, S; Mohsin, KM. "Sirajuddaula" (http://www.banglapedia.org/httpdocs/HT/S_0411.HTM). *Banglapedia*. Asiatic Society of Bangladesh. . Retrieved 2 March 2012.

[19] Fiske, John. "The famine of 1770 in Bengal" (http://etext.library.adelaide.edu.au/f/fiske/john/f54u/chapter9.html). *The Unseen World, and other essays*. Adelaide: University of Adelaide Library Electronic Texts Collection. . Retrieved 26 October 2006.

[20] (Baxter 1997, pp. 30–32)

[21] (Baxter 1997, pp. 39–40)

[22] Wolpert, Stanley (1999). *India* (http://books.google.com/books?id=nHnOERqf-MQC). Berkeley, California, USA: University of California Press. p. 14. ISBN 978-0-520-22172-7. . Retrieved 2 March 2012.

[23] Islam, Sirajul. "Partition of Bengal, 1947" (http://www.banglapedia.org/httpdocs/HT/P_0101.HTM). *Banglapedia*. Asiatic Society of Bangladesh. . Retrieved 3 March 2012.

[24] Dr. Sailen Debnath, 'West Bengal in Doldrums'ISBN 978-81-86860-34-2; & Dr. Sailen Debnath ed. Social and Political Tensions In North Bengal since 1947, ISBN 81-86860-23-1

[25] Dr. Sailen Debnath,ed. Social and Political Tensions In North Bengal since 1947, ISBN 81-86860-23-1.

[26] (Bennett & Hindle 1996, pp. 63–70)

[27] Biswas, Soutik (16 April 2006). "Calcutta's colourless campaign" (http://news.bbc.co.uk/2/hi/south_asia/4909832.stm). BBC. . Retrieved 15 February 2012.

[28] Ghosh Roy, Paramasish (22 July 2005). "Maoist on rise in West Bengal" (http://www.voanews.com/bangla/archive/2005-07/2005-07-22-voa10.cfm). *VOA Bangla*. Voice of America. . Retrieved 11 September 2006.

[29] "Maoist Communist Centre (MCC)" (http://www.satp.org/satporgtp/countries/india/terroristoutfits/MCC.htm). *Left-wing extremist group*. South Asia Terrorism Portal. . Retrieved 11 September 2006.

[30] "Several hurt in Singur clash" (http://www.rediff.com/news/2007/jan/28singur.htm). *rediff news*. 28 January 2007. . Retrieved 15 March 2007.

[31] "Red-hand Buddha: 14 killed in Nandigram re-entry bid" (http://www.telegraphindia.com/1070315/asp/frontpage/story_7519166.asp). *The Telegraph* (Calcutta, India). 15 March 2007. . Retrieved 15 March 2007.

[32] "8 Indian states have more poor than 26 poorest African nations" (http://timesofindia.indiatimes.com/India/8-Indian-states-have-more-poor-than-26-poorest-African-nations/articleshow/6158960.cms). *Times of India* (New Delhi). 22 July 2010. . Retrieved 4 March 2012.

[33] "WB takes the cake when it comes to bandhs" (http://economictimes.indiatimes.com/News/PoliticsNatio/WB_takes_the_cake_when_it_comes_to_bandhs/articleshow/842652.cms). *Economic Times* (New Delhi). 18 December 2006. . Retrieved 4 March 2012.

[34] "Business in West Bengal affected ahead of Tuesday strike" (http://sify.com/finance/business-in-west-bengal-affected-ahead-of-tuesday-strike-news-default-ke0sabdcibh.html). *sify finance*. 26 April 2010. . Retrieved 4 March 2012.

[35] Roy, Hiranmoy; Bhattacharjee, Kaushik (August 2009). "Convergence of human development across Indian States" (http://www.igidr.ac.in/pdf/publication/PP-062-22.pdf) (PDF). Indira Gandhi Institute of Development Research. p. 4. . Retrieved 4 March 2012.

[36] Shah, Mansi (2007). "Waiting for health care: a survey of a public hospital in Kolkata" (http://ccs.in/ccsindia/downloads/intern-papers-08/Waiting-for-Healthcare-A-survey-of-a-public-hospital-in-Kolkata-Mansi.pdf) (PDF). Centre for Civil Society. . Retrieved 31 January 2012.

[37] "West Bengal: health systems development initiative programme memorandum" (http://www.wbhealth.gov.in/Externally_Aided_Projects/HSDI-DFID Programme Memorandum.pdf) (PDF). Government of West Bengal. 15 January 2005. . Retrieved 4 March 2012.

[38] "Impact of social sector development in West Bengal – Midnapore and Birbhum districts" (http://planningcommission.nic.in/reports/sereport/ser/wbm_indx.htm). Planning Commission of India. . Retrieved 4 March 2012.

[39] Mukherjee, Rudrangshu (5 October 2008). "Murder, most foul - the people of Bengal created the darkness that envelops them" (http://www.telegraphindia.com/1081005/jsp/opinion/story_9927371.jsp). *The Telegraph* (Kolkata). . Retrieved 4 March 2012.

[40] "ADB pep pill for Bengal" (http://www.telegraphindia.com/1100613/jsp/business/story_12560050.jsp). *The Telegraph* (Kolkata). 13 June 2010. . Retrieved 4 March 2012.

[41] Ramesh, Randeep (12 November 2007). "Six killed as farmers and communists clash in West Bengal" (http://www.guardian.co.uk/world/2007/nov/12/india.randeepramesh). *The Guardian* (London). . Retrieved 4 March 2012.

[42] "West Bengal political violence continues" (http://economictimes.indiatimes.com/news/politics/nation/West-Bengal-political-violence-continues/articleshow/4871906.cms). *Economic Times* (New Delhi). 8 August 2009. . Retrieved 4 March 2012.

[43] Pal, Supratim (14 May 2007). "Top of world in kingdom of cloud" (http://www.telegraphindia.com/1070514/asp/ranchi/story_7772890.asp). *The Telegraph* (Kolkata). . Retrieved 16 February 2012.

[44] Jayapalan, N (2001). *Foreign policy of India*. New Delhi: Atlantic Publishers and Distributors. p. 344. ISBN 81-7156-898-X.

[45] "Alarming rise in bacterial percentage in Ganga waters" (http://www.thehindubusinessline.in/2006/08/04/stories/2006080402921900.htm). *The Hindu Business Line* (Chennai). 4 August 2006. . Retrieved 4 March 2012.

[46] "Groundwater Arsenic Contamination Status in West Bengal" (http://www.soesju.org/arsenic/wb.htm). *Groundwater Arsenic Contamination in West Bengal – India (17 Years Study)*. School of Environmental Studies, Jadavpur University. . Retrieved October 29, 2006.

[47] "Climate" (http://www.webindia123.com/westbengal/land/climate.htm). *West Bengal: Land*. Suni System (P) Ltd. . Retrieved September 5, 2006.

[48] "kal Baisakhi" (http://amsglossary.allenpress.com/glossary/search?id=kal-baisakhi1). *Glossary of Meteorology*. American Meteorological Society. . Retrieved September 5, 2006.

[49] "State animals, birds, trees and flowers" (http://web.archive.org/web/20090304232302/http://www.wii.gov.in/nwdc/state_animals_tree_flowers.pdf) (PDF). Wildlife Institute of India. Archived from the original (http://www.wii.gov.in/nwdc/state_animals_tree_flowers.pdf) on 4 March 2009. . Retrieved 5 March 2012.

[50] "Forest and tree resources in states and union territories: West Bengal" (http://www.fsi.nic.in/sfr_2009/westbengal.pdf) (PDF). *India state of forest report 2009*. Forest Survey of India, Ministry of Environment & Forests, Government of India. pp. 163–166. . Retrieved 4 March 2012.

[51] "Forest cover" (http://www.fsi.nic.in/sfr_2009/chapter2.pdf) (PDF). *India state of forest report 2009*. Forest Survey of India, Ministry of Environment & Forests, Government of India. pp. 14–24. . Retrieved 4 March 2012.

[52] "Environmental Issues" (http://hdr.undp.org/en/reports/national/asiathepacific/india/India_West Bengal_2004_en.pdf) (PDF). *West Bengal Human Development Report 2004*. Development and Planning Department, Government of West Bengal. May 2004. pp. 180–182. ISBN 81-7955-030-3. . Retrieved 5 March 2012.

[53] Islam, Sadiq (29 June 2001). "World's largest mangrove forest under threat" (http://archives.cnn.com/2001/fyi/student.bureau/06/29/sundarbans/index.html). CNN. . Retrieved 31 October 2006.

[54] Mukherji, S.J. (2000). *College Botany Vol. III: (chapter on Phytogeography)*. Calcutta: New Central Book Agency. pp. 345–365.

[55] "Sundarbans National Park" (http://whc.unesco.org/en/list/452). *World heritage list*. UNESCO World Heritage Center. . Retrieved 4 March 2012.

[56] "Natural vegetation" (http://www.webindia123.com/westbengal/land/forest.htm#N). *West Bengal*. Suni System (P) Ltd. . Retrieved 31 October 2006.

[57] "West Bengal: General Information" (http://web.archive.org/web/20060819094729/http://www.indiainbusiness.nic.in/indian-states/westbengal/General.htm). *India in Business*. Federation of Indian Chambers of Commerce and Industry. Archived from the original (http://www.indiainbusiness.nic.in/indian-states/westbengal/General.htm) on 19 August 2006. . Retrieved 25 August 2006.

[58] "Problems of Specific Regions" (http://hdr.undp.org/en/reports/national/asiathepacific/india/India_West Bengal_2004_en.pdf) (PDF). *West Bengal Human Development Report 2004*. Development and Planning Department, Government of West Bengal. May 2004. pp. 200–203. ISBN 81-7955-030-3. . Retrieved 5 March 2012.

[59] "West Bengal legislative assembly" (http://legislativebodiesinindia.gov.in/States\westbengal\wesbengal-w.htm). *Legislative bodies in India*. National Informatics Centre, India. . Retrieved October 28, 2006.

[60] Delimitation Commission (15 February 2006). "Notification: order no. 18" (http://ceowestbengal.nic.in/news_pdf/gazette123.pdf) (PDF). New Delhi: Election Commission of India. pp. 23–25. . Retrieved 11 February 2012.

[61] "Composition of Rajya Sabha" (http://rajyasabha.nic.in/rsnew/rsat_work/chapter-2.pdf) (PDF). *Rajya Sabha at work*. New Delhi: Rajya Sabha Secretariat. pp. 24–25. . Retrieved 15 February 2012.

[62] "Statewise results - West Bengal" (http://eciresults.nic.in/statewiseS25.htm). Election Commission of India. . Retrieved 13 May 2011.

[63] "Directory of district, sub division, panchayat samiti/ block and gram panchayats in West Bengal, March 2008" (http://www.webel-india.com/blocks n grampanchayats.doc) (DOC). West Bengal Electronics Industry Development Corporation Limited, Government of West Bengal. March 2008. p. 1. . Retrieved 15 February 2012.

[64] "Section 2 of West Bengal Panchayat Act, 1973" (http://wbdemo5.nic.in/html/asp/g2csw/sections/2.htm). Department of Panchayat and Rural Department, West Bengal. . Retrieved December 9, 2008.

[65] "Urban agglomerations/cities having population 1 million and above" (http://censusindia.gov.in/2011-prov-results/paper2/data_files/india2/Million_Plus_UAs_Cities_2011.pdf). *Provisional population totals, census of India 2011*. The Registrar General & Census Commissioner, India. 2011. . Retrieved 26 January 2012.

[66] "Cities having population 1 lakh and above, census 2011" (http://www.censusindia.gov.in/2011-prov-results/paper2/data_files/India2/Table_2_PR_Cities_1Lakh_and_Above.pdf). *Provisional population totals, census of India 2011*. The Registrar General & Census Commissioner, India. . Retrieved 18 October 2011.

[67] "West Bengal" (http://www.ibef.org/download/West_Bengal_271211.pdf). India Brand Equity Foundation. November 2011. . Retrieved 6 February 2012.

[68] "Industrial infrastructure" (http://www.wbidc.com/about_wb/industrial_infrastructure.htm). West bengal Industrial Development Corporation. . Retrieved 5 March 2012.

[69] "Introduction and Human Development Indices for West Bengal" (http://hdr.undp.org/en/reports/national/asiathepacific/india/India_West Bengal_2004_en.pdf) (PDF). *West Bengal Human Development Report 2004*. Development and Planning Department, Government of West Bengal. May 2004. pp. 4–6. ISBN 81-7955-030-3. . Retrieved 5 March 2012.

[70] "Mamata seeks debt restructuring plan for West Bengal" (http://articles.economictimes.indiatimes.com/2011-10-22/news/30309832_1_debt-restructuring-plan-debt-burden-12th-plan). *Economic Times* (New Delhi). 22 October 2011. . Retrieved 4 March 2012.

[71] Ray Choudhury, Ranabir (27 October 2006). "A new dawn beckons West Bengal" (http://www.thehindubusinessline.com/2006/10/27/stories/2006102700080100.htm). *The Hindu Business Line* (Chennai). . Retrieved 29 October 2006.

[72] "West Bengal Industrial Development Corporation Ltd." (http://www.indiaathannover.org/pdf/exhibitorslist.pdf) (PDF). *India @ Hannover Messe 2006*. Engineering Export Promotion Council (EEPC), India. p. 303. . Retrieved September 7, 2006.

[73] Saha, Sambit (9 September 2003). "Nathula trade may spur business in NE" (http://www.rediff.com/money/2003/sep/09trading.htm). rediff.com. . Retrieved 18 September 2007.

[74] Raja Mohan, C. (16 July 2004). "A foreign policy for the East" (http://www.hindu.com/2004/07/16/stories/2004071601841000.htm). Chennai: The Hindu. . Retrieved 5 March 2012.

[75] "Statewise Length of national highways in India" (http://morth.nic.in/showfile.asp?lid=366). *National Highways*. Department of Road Transport and Highways; Ministry of Shipping, Road Transport and Highways; Government of India. . Retrieved 9 February 2012.

[76] Chattopadhyay, Suhrid Sankar (January–February 2006). "Remarkable Growth" (http://www.flonnet.com/fl2302/stories/20060210004209800.htm). *Frontline* (Chennai, India: The Hindu) **23** (2). . Retrieved March 31, 2008.

[77] "Geography : Railway Zones" (http://www.irfca.org/faq/faq-geog.html). *IRFCA.org*. Indian Railways Fan Club. . Retrieved August 31, 2007.

[78] "About Kolkata Metro" (http://www.kolmetro.com/). Kolkata Metro. . Retrieved September 1, 2007.

[79] "Mountain Railways of India" (http://whc.unesco.org/en/list/944). UNESCO World Heritage Centre. . Retrieved April 30, 2006.

[80] "Port info: cargo statistics" (http://www.kolkataporttrust.gov.in/). *Kolkata Port Trust*. Kolkata Port Trust, India. . Retrieved 9 February 2012.

[81] "Intra-city train travel" (http://timesfoundation.indiatimes.com/articleshow/657741.cms). *reaching India* (Times Internet Limited). . Retrieved August 31, 2007.

[82] "India ferry disaster kills scores" (http://www.bbc.co.uk/news/world-south-asia-11679012). *BBC News*. November 2, 2010. .

[83] "Census Population" (http://indiabudget.nic.in/es2006-07/chapt2007/tab97.pdf) (PDF). *Census of India*. Ministry of Finance India. . Retrieved December 18, 2008.

[84] Hoddie, Matthew (2006). *Ethnic realignments: a comparative study of government influences on identity* (http://books.google.com/books?id=6ka0nMJgKbYC). Lexington Books. pp. 114–115. ISBN 978-0-7391-1325-7. . Retrieved 16 February 2012.

[85] Banerjee, Himadri; Gupta, Nilanjana; Mukherjee, Sipra, eds. (2009). *Calcutta mosaic: essays and interviews on the minority communities of Calcutta* (http://books.google.com/books?id=cSTEOx_Lw9MC&dq). Anthem Press. p. 3. ISBN 978-81-905835-5-8. . Retrieved 29 January 2012.

[86] Banerjee, Himadri; Gupta, Nilanjana; Mukherjee, Sipra, eds. (2009). *Calcutta mosaic: essays and interviews on the minority communities of Calcutta* (http://books.google.com/books?id=cSTEOx_Lw9MC&dq). Anthem Press. pp. 9–10. ISBN 978-81-905835-5-8. . Retrieved 29 January 2012.

[87] "Data on Religion" (http://web.archive.org/web/20070812142520/http://www.censusindia.net/religiondata/). *Census of India (2001)*. Office of the Registrar General & Census Commissioner, India. Archived from the original (http://www.censusindia.gov.in/) on August 12, 2007. . Retrieved August 26, 2006.

[88] "Report of the Commissioner for linguistic minorities: 47th report (July 2008 to June 2010)" (http://nclm.nic.in/shared/linkimages/NCLM47thReport.pdf). Commissioner for Linguistic Minorities, Ministry of Minority Affairs, Government of India. pp. 122–126. . Retrieved 16 February 2012.

[89] Population of West Bengal (80,221,171) is 7.8% of India's population (1,027,015,247)

[90] "Table 1: Distribution of population, sex ratio, density and decadal growth rate of population: 2011" (http://www.censusindia.gov.in/2011-prov-results/prov_results_paper1_india.html). *Provisional population totals paper 1 of 2011 India: series 1*. Registrar General & Census Commissioner, India. . Retrieved 16 February 2012.

[91] "Table 2(3): Literates and literacy rates by sex : 2011" (http://www.censusindia.gov.in/2011-prov-results/prov_results_paper1_india.html). *Provisional population totals paper 1 of 2011 India: series 1*. Registrar General & Census Commissioner, India. . Retrieved 16 February 2012.

[92] "An Indian life: Life expectancy in our nation" (http://www.indiatogether.org/health/infofiles/life.htm). *India Together*. Civil Society Information Exchange Pvt. Ltd. . Retrieved August 26, 2006.

[93] National Crime Records Bureau (2004). "Crimes in Mega Cities" (http://ncrb.nic.in/crime2004/cii-2004/CHAP2.pdf) (PDF). *Crime in India-2004*. Ministry of Home Affairs. p. 158. . Retrieved August 26, 2006.

[94] "Human security" (http://hdr.undp.org/en/reports/national/asiathepacific/india/India_West Bengal_2004_en.pdf) (PDF). *West Bengal Human Development Report 2004*. Development and Planning Department, Government of West Bengal. May 2004. pp. 167–172. ISBN 81-7955-030-3. . Retrieved 5 March 2012.

[95] National Crime Records Bureau (2004). "General Crime Statistics Snapshots 2004" (http://ncrb.nic.in/crime2004/cii-2004/Snapshots.pdf) (PDF). *Crime in India-2004*. Ministry of Home Affairs. p. 1. . Retrieved April 26, 2006.

[96] "Plight of prostitutes in Kolkata" (http://www.merinews.com/article/plight-of-prostitutes-in-kolkata/137536.shtml). Merinews.com. . Retrieved 2012-01-09.

[97] "The Bauls of Bengal" (http://bengalonline.sitemarvel.com/bengali-folklore.asp?art=baul). *Folk Music*. BengalOnline. . Retrieved October 26, 2006.

[98] "Chau: The Rare Mask Dances" (http://www.boloji.com/dances/00109.htm). *Dances of India*. Boloji.com. . Retrieved October 22, 2006.

[99] Gertjan de Graaf, Abdul Latif. "Development of freshwater fish farming and poverty alleviation: A case study from Bangladesh" (http://govdocs.aquake.org/cgi/reprint/2003/1201/12010300.pdf) (PDF). Aqua KE Government. . Retrieved October 22, 2006.

[100] Saha, S (January 18, 2006). "Resurrected, the kathi roll – Face-off resolved, Nizam's set to open with food court" (http://www.telegraphindia.com/1060118/asp/calcutta/story_5733258.asp). Calcutta, India: The Telegraph (Kolkata). . Retrieved October 26, 2006.

[101] "Mobile food stalls" (http://www.bangalinet.com/mobile_foodstalls.htm). Bangalinet.com. . Retrieved October 26, 2006.

[102] "Durga Puja" (http://www.westbengaltourism.gov.in/web/guest/festival-home). *Festivals celebrated throughout West Bengal*. Department of Tourism, Government of West Bengal. . Retrieved 5 March 2012.

[103] "Boards of secondary & senior secondary education in India" (http://mhrd.gov.in/recognized_boards). Department of School Education and Literacy, Ministry of Human Resource Development, Government of India. . Retrieved 18 April 2012.

[104] "UGC recognised Universities in West Bengal with NAAC accreditation status" (http://www.educationobserver.com/resources/universsities/west_bengal.htm). Education Observer. . Retrieved October 26, 2006.

[105] "West Bengal University of Health Sciences" (http://www.thewbuhs.org/). West Bengal University of Health Sciences. . Retrieved October 26, 2006.

[106] "List of Affiliated Colleges" (http://web.archive.org/web/20080201164051/http://www.caluniv.ac.in/coll.htm). University of Calcutta. Archived from the original (http://www.caluniv.ac.in/coll.htm) on February 1, 2008. . Retrieved March 29, 2008.

[107] Mitra, P (August 31, 2005). "Waning interest" (http://www.telegraphindia.com/1050831/asp/careergraph/story_5174502.asp). *Careergraph* (Calcutta, India: The Telegraph). . Retrieved October 26, 2006.

[108] "Visva-Bharati: Facts and Figures at a Glance" (http://www.visva-bharati.ac.in/at_a_glance/at_a_glance.htm). Visva-Bharati Computer Centre. . Retrieved March 31, 2007.

[109] "General Review" (https://rni.nic.in/pii.asp). Registrar of Newspapers for India. . Retrieved 6 March 2012.

[110] "Bengali News Channel took 5 months to reach no.1 position" (http://www.moneycontrol.com/news/business/bengali-news-channel-took-5-months-to-reach-no1-position_242437.html). News Center. . Retrieved Sep 7, 2006.

[111] "CALCUTTA : Television, Radio Channels" (http://www.calcuttaweb.com/tvradio.shtml). Calcutta Web. . Retrieved Sep 7, 2006.

[112] Prabhakaran, Shaji (January 18, 2003). "Football in India – A Fact File" (http://www.longlivesoccer.com/indiafootball.htm). LongLiveSoccer.com. . Retrieved October 26, 2006.

[113] "Indian Football Clubs" (http://www.iloveindia.com/sports/football/clubs/index.html). Iloveindia.com. . Retrieved October 26, 2006.

[114] "History of Polo" (http://www.hpa-polo.co.uk/about/history_polo.asp). Hurlingham Polo Association. . Retrieved August 30, 2007.

[115] "Royal Calcutta Golf Club" (http://www.britannica.com/eb/topic-511285/Royal-Calcutta-Golf-Club). Encyclopaedia Britannica. . Retrieved August 30, 2007.

[116] "India – Eden Gardens (Kolkata)" (http://www.cricketweb.net/country/venue.php?CategoryIDAuto=12&VenueIDAuto=26). Cricket Web. . Retrieved October 26, 2006.

[117] "100 000+ Stadiums" (http://www.worldstadiums.com/stadium_menu/stadium_list/100000.shtml). World Stadiums. . Retrieved October 26, 2006.

[118] "The Asian Football Stadiums (30.000+ capacity)" (http://www.fussballtempel.net/afc/listeafc.html). Gunther Lades. . Retrieved October 26, 2006.

[119] Raju, Mukherji (March 14, 2005). "Seven Years? Head Start" (http://www.telegraphindia.com/1050314/asp/opinion/story_4428341.asp). Calcutta, India: The Telegraph. . Retrieved October 26, 2006.

[120] "Sports & Adventure" (http://www.wbtourism.com/sports_adventure/index.htm). West Bengal Tourism. . Retrieved October 22, 2006.

[121] "Famous Indian Football Players" (http://www.iloveindia.com/sports/football/players/index.html). Iloveindia.com. . Retrieved October 26, 2006.

References

- Baxter, C (1997). *Bangladesh, From a Nation to a State*. Westview Press. p. 0813336325. ISBN 1-85984-121-X

- Bennett, A; Hindle, J (1996). *London Review of Books: An Anthology*. Verso. pp. 63–70. ISBN 1-85984-121-X

- Roy, A; Alsayyad, N (2004). *Urban Informality: Transnational Perspectives from the Middle East, Latin America and South Asia*. Lexington Books. ISBN 0-7391-0741-0

- *West Bengal Human Development Report 2004* (http://hdr.undp.org/en/reports/national/asiathepacific/india/India_West Bengal_2004_en.pdf). Development and Planning Department, Government of West Bengal. 2004. ISBN 81-7955-030-3

- *Impact of Social Sector Development in West Bengal* (http://planningcommission.nic.in/reports/sereport/ser/wbm_indx.htm). Planning Commission, Government of India. 2009.

- Klass, L; Morton, S (1996). *Community Structure and industrialization in West Bengal*. University Press of America Inc.. ISBN 0-7618-0420-X

- Sunny, C (1999). "Poverty and social development in west bengal" (http://planningcommission.nic.in/reports/sereport/ser/wbm/wbm_ch2.pdf). *India Rural Development Report, NIRD*. Retrieved 1999

- KPMG India, V (December 10, 2001). "Sustainable economic development in West Bengal – A Perspective" (http://www.in.kpmg.com/TL_Files/Pictures/West_Bengal.pdf). *Confederation of Indian Industry (CII)*. Retrieved 2007

- Amrita Basu, V. (1997). *Two Faces of Protest: Contrasting Modes of Women's Activism in India* (http://books.google.com/?id=ZyY0Yb5BrqgC&pg=PA25&dq=communism+in+west+bengal&cd=5#v=onepage&q=communism in west bengal). University of California Press ltd.. ISBN 0-520-06506-9. Retrieved June 16, 2009.

- Jasodhara Bagchi, Sarmistha Dutta Gupta, V. (2000). *The changing status of women in West Bengal, 1970–2000: the challenge ahead* (http://books.google.com/?id=KYYW8Un5zFAC&pg=PA119&dq=violence+west+bengal&cd=1#v=onepage&q=violence west bengal). Saga Publication India Pvt Ltd.. ISBN 0-7619-3242-9. Retrieved June 16, 2010.

- Magnus Öberg, Kaare Strom, V. (2008). *Resources, governance and civil conflict* (http://books.google.com/?id=eBW-KtJ28ZsC&pg=PA93&dq=Naxalite+in+west+bengal&cd=1#v=onepage&q=Naxalite in west bengal). Routledge. ISBN 978-0-415-41671-9. Retrieved June 16, 2004.

- Atul Kohli, I. (1987). *The State and Poverty in India* (http://books.google.com/?id=vxLAK8EXo84C&pg=PA117&dq=poverty+in+west+bengal&cd=1#v=onepage&q= west bengal). Cambridge University Press. ISBN 978-0-521-37876-5. Retrieved June 16, 2007.

- Marvin, Davis (1983) [1983]. *Rank and rivalry: the politics of inequality in rural West Bengal*. Cambridge: Cambridge University Press. ISBN 0-521-24657-1.
- Richard Maxwell Eaton, The rise of Islam and the Bengal frontier, 1204–1760, 1993, University of California Press, California, California,1993, ISBN 0-520-08077-7.
- Ross Mallick. (1955). Development Policy of a Communist Government: West Bengal Since 1977, Cambridge University Press, Cambridge (Reprinted 2008) ISBN 978-0-521-43292-4.
- Jasodhara Bagchi, Sarmistha Dutta Gupta, V. (2009). *A Story of Ambivalent Modernization in Bangladesh and West Bengal: The Rise and Fall of Bengali Elitism in South Asia*. Peter Lang Publishing; First printing edition. ISBN 978-1-4331-0820-4.
- Tapan Raychaudhuri (2002). *Europe Reconsidered: Perceptions of the West in Nineteenth-Century Bengal*. Oxford University Press. ISBN 978-0-19-566109-5.
- Harriss-White, Barbara (editor) (2008). *Rural Commercial Capital: Agricultural Markets in West Bengal*. Oxford University Press, USA. ISBN 0-19-569159-8.
- Raychaudhuri, Ajitava (editor); Das, Tuhin K. (editor) (2005). *West Bengal economy: some contemporary issues* (http://books.google.com/?id=NTeHPuhTsXcC&pg=PA45&dq=politics+in+west+bengal& cd=42#v=onepage&q). Jadavpur University Press, India. ISBN 81-7764-731-8.
- Inden; Ronald B.; Ralph W (2005). *Kinship in Bengali Culture*. The University of Chicago Press, 1977. ISBN 81-8028-018-7.
- Davis, Marvin (1983). *Rank and rivalry: the politics of inequality in rural West Bengal*. 1st edition. Cambridge University Press. xxvii, 239. ISBN 0-521-24657-1.
- Banerjee, Anuradha (1998). *Environment, population, and human settlements of Sundarban Delta*. Ashok Kumar Mittal. ISBN 81-7022-739-9.

External links

Government

- Official West Bengal Government Web Portal (http://www.westbengal.gov.in/)
- Department of Tourism, Government of West Bengal (http://www.westbengaltourism.gov.in/web/guest/index)
- Directorate of Commercial Taxes, Government of West Bengal (http://www.wbcomtax.nic.in/welcome.asp)
- West Bengal Information Commission (http://wbic.gov.in/)

Other

- West Bengal travel guide from Wikitravel
- West Bengal (http://www.britannica.com/EBchecked/topic/640088/West-Bengal) *Encyclopædia Britannica* entry
- West Bengal (http://www.dmoz.org/Regional/Asia/India/West_Bengal/) at the Open Directory Project
- West Bengal (http://www.wikimapia.org/#lat=22.5697&lon=88.3697&z=10&l=0&m=b) Satellite view at WikiMapia

Article Sources and Contributors

Budge_Budge_I_(community_development_block) *Source*: http://en.wikipedia.org/w/index.php?title=Budge_Budge_I_%28community_development_block%29 *Contributors*: Chandan Guha

Alipore_Sadar_subdivision *Source*: http://en.wikipedia.org/w/index.php?title=Alipore_Sadar_subdivision *Contributors*: Chandan Guha, GDibyendu, Jayantanth, LilHelpa, Mhchintoo, Paalappoo, Sadads, Skier Dude, 1 anonymous edits

South_24_Parganas_district *Source*: http://en.wikipedia.org/w/index.php?title=South_24_Parganas_district *Contributors*: Antorjal, Bhadani, Certes, Chandan Guha, Chhoton, Chris the speller, Crusoe8181, Debasishkoley, FourthAve, GDibyendu, Generalboss3, Jayantanth, Joy1963, Kwamikagami, LilHelpa, Lucio Di Madaura, MK2, Materialscientist, P.K.Niyogi, PhnomPencil, Sadads, Simplyanon, Tabletop, Tassedethe, WoodElf, अमित भटनागर, , 30 anonymous edits

Budge_Budge *Source*: http://en.wikipedia.org/w/index.php?title=Budge_Budge *Contributors*: Antorjal, Ardric47, BD2412, Bhadani, Chandan Guha, Crusoe8181, D6, Dwaipayanc, EoGuy, Euchiasmus, Fayenatic london, GDibyendu, Gaius Cornelius, Gilliam, Husond, Ixfd64, Jethwarp, LilHelpa, Ling.Nut, MER-C, Mild Bill Hiccup, P.K.Niyogi, Paalappoo, Parthab96, Rich Farmbrough, Ritroy, Rjwilmsi, SameerKhan, Sardanaphalus, Sfan00 IMG, Shantanu shukla, Stepheng3, Swestrup, Tabletop, 65 anonymous edits

Balarampur,_Budgebudge *Source*: http://en.wikipedia.org/w/index.php?title=Balarampur%2C_Budgebudge *Contributors*: Chandan Guha, D6, GDibyendu, MozzyMo, P.K.Niyogi

Community_development_block_in_India *Source*: http://en.wikipedia.org/w/index.php?title=Community_development_block_in_India *Contributors*: Ageo020, Ansumang, Beyond My Ken, Bogdan Nagachop, Chandan Guha, Explicit, JDP90,

Gram_panchayat *Source*: http://en.wikipedia.org/w/index.php?title=Gram_panchayat *Contributors*: AdjustShift, Ageo020, Alpha Quadrant (alt), AndrewRT, Andycjp, Angers roams, Apokrif, Auntof6, Avoided, Beyond My Ken, Bittupanpaliya, Bushcarrot, CalJW, Chrisminter, Cybercobra, DBigXray, Dbachmann, Dev, DhruvDhamani, Faradayplank, Feezo, Fieldday-sunday, GTBacchus, Indu, Jackerhack, James500, Ligulem, Loqugr, Lucio Di Madaura, MKar, Nichalp, Nisrec, Phlyght, Priyatu, Qst, Rajankila, Rama's Arrow, Sailing Stonehenge, Salilb, Scopecreep, Shadowjams, Shyamsunder, Spartiate, Spencer, Sskoslia786, Starrahul, Tabletop, ThomasK, Tommy2010, Uncle G, Utcursch, Vprajkumar, WarFox, 87 anonymous edits

Scheduled_castes_and_scheduled_tribes *Source*: http://en.wikipedia.org/w/index.php?title=Scheduled_castes_and_scheduled_tribes *Contributors*: Agsrivaths, Alai, Allen4names, Almithra, Altenmann, Ambuj.Saxena, Andrew Gwilliam, Andycjp, Anwar saadat, Arjun024, Ashokkw, Atul903, Austria156, Babbage, Bigbrothersorder, Borfee, BrokenSegue, CJLL Wright, CarTick, Cdamama, Chancemill, ChrisHodgesUK, Clarkpoon, Cpeel, Dbachmann, Dlempa, Doc Tropics, Dr.K., Dragon guy, Drmies, Dwaipayanc, Earthlyreason, Editor2020, Ekabhishek, Emilio Juanatey, FolkTraditionalist, George Burgess, GeorgeLouis, Gotipe, Hkelkar, Humanrights2011, Humboldt, Idleguy, Indianpeacock, Int21h, J.delanoy, Jhartmann, Johari Shauka, JohnLobster, Kaihsu, Kanatonian, Kpalion, LeeHunter, Lhsrhsbvs, Liftarn, Logrithm, Mandarax, Mark Arsten, MatthewVanitas, Mattisse, Mel Etitis, Michael Hardy, Mkrestin, Network for social accountability, Nichalp, Nnemo, Nriweb, Piotrus, Pol430, Prashantroy, Prodiptodas123, Protozoan, QuartierLatin1968, Randhirreddy, Reason turns rancid, Samathawiki, Satendrav80, Sdeepak scor, Shaeed baba Sangat singh ji, SimonP, Sitush, Snowolf, Steinrog2, Tainter, Teleutomyrmex, Truthseeker81, Unnikn, UplinkAnsh, Utcursch, Valentinejoesmith, Valrama, Vinodmikkili, Vkrohit rohit, Wasbeer, Will Beback, Woohookitty, Ybbor, Yogendra Uikey, 167 anonymous edits

West_Bengal *Source*: http://en.wikipedia.org/w/index.php?title=West_Bengal *Contributors*: *drew, 5 albert square, Aaroncrick, Abdussamad77, Abhisek2091, Abhowmick, Acsenray, Admrboltz, AgnosticPreachersKid, Ahoerstemeier, Airunp, AjaxSmack, Ak2431989, Akkida, Al Silonov, AlexanderKaras, AlimanRuna, Alokchakrabarti, Aloke Kumar, Amartyabag, Ambuj.Saxena, Amitabdev, Amitsanyal, Amplitude101, Anchitk, Anclation, Andrewlp1991, Andyindia, Aneesharry, Aniketmg, Anirvan, Ankitbhatt, Annalise, Anshuman.jrt, Antorjal, Anwar saadat, Aotearoa, AralSea, Arnab kl4, Arnabik, Arnabik9, Arparag, Art LaPella, Aryasanyal, Ashowmega, Astrokid, Aurorion, Avenue X at Cicero, AvicAWB, Avinashv11, Avr0716ap, AxelBoldt, BD2412, Babuonwiki, Bapan25, Bbbl67, Begoon, Bharatveer, Biblbroks, Bijoy Krishna Das, Bility, Bl4ze013, Bless sins, Blue-Haired Lawyer, Bobblewik, Bonadea, BostonMA, Brewcrewer, Brighterorange, Buaidh, CJLL Wright, CLW, Cab.jones, Cactus.man, Caeruleancentaur, Caltas, CarTick, Cecil, Chandan Guha, Charles Matthews, Charlesdrakew, Chirag, Chrism, Cloudbound, Clt13, Combustion X, Conversion script, D, D6, DARTH SIDIOUS 2, DMS, DaGizza, Dark Shikari, Dasoumendu, Dav subrajathan.357, David matthews, Dawn Bard, Debasishkoley, Debjato, Deepak D'Souza, Deepika11, Deeptrivia, Dekimasu, Der Golem, Dewan357, Dipankan001, Discospinster, Dlohcierekim, Dn9ahx, Domino theory, Drmies, Drpickem, Drumguy8800, Dwaipayanc, East Bahamas, Ed Poor, El C, Elockid, Eluchil404, Enti342, Enviroboy, EoGuy, Epipelagic, Eras-mus, Eukesh, Expresswaytoparadise, Filemon, Fnielsen, Fratrep, Fundamental metric tensor, GDibyendu, GSMR, Gaius Cornelius, Galoubet, Ganeshk, Gangulybiswarup, Gaurav38, Gene Nygaard, Generalboss3, Gimmetrow, Gman124, Goatse Rulz, Gogo Dodo, GoingBatty, Golbez, Gomada, Good Olfactory, Gopalagarwal11, Gppande, Graham87, Grenavitar, Grey Scaler, Ground Zero, Gurubrahma, Gzornenplatz, Hagedis, Haham hanuka, HeBhagawan, Hello2abir, Help Always, Hemanshu, Hindubengali, Hippietrail, Holy Ganga, Hometech, Hqb, Ian Pitchford, Improv, InMooseWeTrust, India Gate, Indon, Indranee, Interchange88, JForget, JaGa, Jac16888, Jagged 85, Jahangard, Japanese Searobin, Jauhienij, Jayantanth, Jeroje, Jethwarp, Jiminy900, Jimp, Jmgarg1, Jojit fb, Jon4562002, JonnyNYC90, Jonoikobangali, Jorunn, Jovianeye, Joy9, Joyson Prabhu, Justarijit, Kamalakardandu, Kannanwrites, Kartik2008, Kaushikhm, Kayau, Kbdank71, Kensplanet, Khazar, Kintetsubuffalo, Kiril Simeonovski, Kirti 1102, Kkm010, KnowledgeHegemony, Knutux, Koavf, Kozuch, Kristian Vangen, Kuaichik, Kungming2, Kurykh, KuwarOnline, Kwamikagami, Lexington1, Lightmouse, Ligulem, Livajo, Lkinkade, Logan, Logicalthinker33, Lollywood, Lord of the Goatse, LordGulliverofGalben, LordSimonofShropshire, LordSuryaofShropshire, MJCdetroit, Maddyr, Mainak, Mani1, Manoj nav, Martial75, Master Of Ninja, Materialscientist, Mattbr, Maurice45, Mayukh iitbombay 2008, Mboverload, Mddeath, MeekSaffron, MegaSloth, Mhchintoo, Michael Devore, Michaelmas1957, Mightymights, Mike R, Mike Rosoft, MikeLynch, Mild Bill Hiccup, Millahnna, Minusia, Mister Goat, Mkeane1019, Mkweise, Morwen, Mr Tan, Mr0t1633, Munci, Munita Prasad, Murphy dean, Murtasa, Mvp15, My76Strat, Nafsadh, Nakon, Naru12333, Nat Krause, Nataliehorsey, Nataraja, NawlinWiki, Neddyseagoon, NeelAbodh, Nichalp, Nigholith, Nikkul, Niteowlneils, Niteshpradhans, Northumbrian, Npeters22, Nunamiut, Obradovic Goran, OhanaUnited, Olivier, OneGuy, Oolong, Overvital1288, Oystertoadfish, P.K.Niyogi, Paleolithic1288, Patel24, Patrick, Paul-L, Pavel Vozenilek, Pax:Vobiscum, Pdutta81, Peoplespower, Petiatil, Pgk, Philyiverson3, PhnomPencil, PigFlu Oink, Piledhigheranddeeper, Pill, Pit, Planemad, Pollodiablowiki, Ppntori, Pradip ch, Pradiptaray, Priyatu, Prodego, Proofreader77, Proud Ho, Psubhashish, Pupunwiki, QuartierLatin1968, Quibik, Qwerta369, Qxz, R'n'B, RJFJR, Rabbabodrool, Radar signal, Ragib, Rahulghose, Raj Krishnamurthy, Rama's Arrow, Ramayan, Raul654, Rbha7, Recognizance, Redtigerxyz, Rehan.haider, RexNL, Riana, Rich Farmbrough, Rick Block, Ritwikbmca, Rjwilmsi, Robertgreer, Ronline, Rudroanik, Rueben lys, Saga City, Saimdusan, Sam Hocevar, SameerKhan, Samir, Sander123, SandyGeorgia, Saptarshivet, Saravask, Sardanaphalus, Sarvagnya, Saurabh.vinian, Savitr, Sayanmitra, Scjessey, Scope creep, Scottinglis, ScribeOO, Scriber, Sd31415, Sen.tanima, Shalimer, Shmitra, Siddhartha Ghai, Skier Dude, Skinsmoke, Skylark2008, Skylerb, Sluzzelin, Slysplace, Snthakur, Soman, SomeStranger, Soumitrahazra, Souravdefermat, SpacemanSpiff, Sport woman, Sproy, Spundun, Src2206, Srich32977, Srikeit, Srini81, Stepheng3, Storkk, Subhashischakraborty, Sudipta.kamila, Sukanya92, Sumanch, Sunday Sunny Gill265, Super cyclist, Supray, Supten, Surajcap, Sushant gupta, Swpb, Taajikhan, Tamilan101, Taratv, Tbhotch, Tbone, Technopilgrim, Template namespace initialisation script, Terissn, Tesi1700, Thaejas, The Magnificent Clean-keeper, The Thing That Should Not Be, Thisthat2011, Thunderboltz, TimBentley, Timberframe, TobyDZ, Tom Edwards, Tom Radulovich, Tony1, Tpbradbury, Trakesht, Travelbird, Treisijs, Trinanjon, Troglo, Tuncrypt, Tutmosis, UnicornTapestry, Unmesh Bangali, Urnonav, UrsusArctosL71, VT hawkeye, Vald, Vale of Glamorgan, Vedran12, Vgranucci, Victor D, Vijayaditya, Wackymacs, Wareq, Wavelength, Whitejay251, Wik, Wiki-uk, Winston365, WoodElf, Woohookitty, World8115, Wouterhagens, Wtmitchell, Wywhpf, Xufanc, Yamamoto Ichiro, Yarnalgo, YellowMonkey, Yorke.aaron, Zakuragi, Zbd, Zeman, Ziaur, Zundark, , , 538 anonymous edits

Image Sources, Licenses and Contributors

file:India West Bengal location map.svg *Source*: http://en.wikipedia.org/w/index.php?title=File:India_West_Bengal_location_map.svg *License*: unknown *Contributors*: NordNordWest, 1 anonymous edits

File:Locator_Dot.svg *Source*: http://en.wikipedia.org/w/index.php?title=File:Locator_Dot.svg *License*: unknown *Contributors*: Petr Dlouhý

File:Flag of India.svg *Source*: http://en.wikipedia.org/w/index.php?title=File:Flag_of_India.svg *License*: unknown *Contributors*: Anomie, Mifter

file:India West Bengal locator map.svg *Source*: http://en.wikipedia.org/w/index.php?title=File:India_West_Bengal_locator_map.svg *License*: unknown *Contributors*: Planemad, 2 anonymous edits

file:Red pog.svg *Source*: http://en.wikipedia.org/w/index.php?title=File:Red_pog.svg *License*: unknown *Contributors*: Anomie

Image:South 24 Parganas district.svg *Source*: http://en.wikipedia.org/w/index.php?title=File:South_24_Parganas_district.svg *License*: unknown *Contributors*: GDibyendu

Image:Setup of India.png *Source*: http://en.wikipedia.org/w/index.php?title=File:Setup_of_India.png *License*: unknown *Contributors*: User:V Prajkumar

File:India West Bengal locator map.svg *Source*: http://en.wikipedia.org/w/index.php?title=File:India_West_Bengal_locator_map.svg *License*: unknown *Contributors*: Planemad, 2 anonymous edits

File:West Bengal locator map.svg *Source*: http://en.wikipedia.org/w/index.php?title=File:West_Bengal_locator_map.svg *License*: unknown *Contributors*: Nichalp, Planemad, 1 anonymous edits

File:Decrease2.svg *Source*: http://en.wikipedia.org/w/index.php?title=File:Decrease2.svg *License*: unknown *Contributors*: User:Sarang

Image:Pala Empire (Dharmapala).gif *Source*: http://en.wikipedia.org/w/index.php?title=File:Pala_Empire_(Dharmapala).gif *License*: unknown *Contributors*: User:nawab_of_dhaka

Image:Devapala.jpg *Source*: http://en.wikipedia.org/w/index.php?title=File:Devapala.jpg *License*: unknown *Contributors*: Nawab of dhaka

File:Raja Ram Mohan Roy.jpg *Source*: http://en.wikipedia.org/w/index.php?title=File:Raja_Ram_Mohan_Roy.jpg *License*: unknown *Contributors*: Original uploader was Rueben lys at en.wikipedia

File:Subhas Bose.jpg *Source*: http://en.wikipedia.org/w/index.php?title=File:Subhas_Bose.jpg *License*: unknown *Contributors*: Original uploader was Sidha at en.wikipedia

File:Kalinagar Floods B.JPG *Source*: http://en.wikipedia.org/w/index.php?title=File:Kalinagar_Floods_B.JPG *License*: unknown *Contributors*: Original uploader was Soumyasch at en.wikipedia

File:Teestavalley.jpg *Source*: http://en.wikipedia.org/w/index.php?title=File:Teestavalley.jpg *License*: unknown *Contributors*: Mircea, Nichalp

File:Fishing Cat (Prionailurus viverrinus) 3.jpg *Source*: http://en.wikipedia.org/w/index.php?title=File:Fishing_Cat_(Prionailurus_viverrinus)_3.jpg *License*: unknown *Contributors*: Cliff

File:White-throated Kingfisher (Shankar).jpg *Source*: http://en.wikipedia.org/w/index.php?title=File:White-throated_Kingfisher_(Shankar).jpg *License*: unknown *Contributors*: AshLin, Bidgee, Conscious, Keith Edkins, Leoboudv, Liné1, MBisanz, MPF

File:Alstonia scholaris.jpg *Source*: http://en.wikipedia.org/w/index.php?title=File:Alstonia_scholaris.jpg *License*: unknown *Contributors*: User:AmarChandra

File:Flower & flower buds I IMG 2257.jpg *Source*: http://en.wikipedia.org/w/index.php?title=File:Flower_&_flower_buds_I_IMG_2257.jpg *License*: unknown *Contributors*: User:J.M.Garg

Image:Panthera tigris tigris.jpg *Source*: http://en.wikipedia.org/w/index.php?title=File:Panthera_tigris_tigris.jpg *License*: unknown *Contributors*: User:Zwoenitzer

Image:Arabari.jpg *Source*: http://en.wikipedia.org/w/index.php?title=File:Arabari.jpg *License*: unknown *Contributors*: Antorjal at en.wikipedia

File:Calcutta High Court.jpg *Source*: http://en.wikipedia.org/w/index.php?title=File:Calcutta_High_Court.jpg *License*: unknown *Contributors*: Herzi Pinki, Roland zh

File:WestBengalDistricts numbered.svg *Source*: http://en.wikipedia.org/w/index.php?title=File:WestBengalDistricts_numbered.svg *License*: unknown *Contributors*: Original uploader was Antorjal at en.wikipedia Later version(s) were uploaded by Deeptrivia at en.wikipedia.

File:Vegetable vendor.JPG *Source*: http://en.wikipedia.org/w/index.php?title=File:Vegetable_vendor.JPG *License*: unknown *Contributors*: User:Gangulybiswarup

File:PaddyandjuteBengal.JPG *Source*: http://en.wikipedia.org/w/index.php?title=File:PaddyandjuteBengal.JPG *License*: unknown *Contributors*: Dwaipayanc

Image:Indian Rupee symbol.svg *Source*: http://en.wikipedia.org/w/index.php?title=File:Indian_Rupee_symbol.svg *License*: unknown *Contributors*: User:Orionist

File:KolkataLocalTrains.JPG *Source*: http://en.wikipedia.org/w/index.php?title=File:KolkataLocalTrains.JPG *License*: unknown *Contributors*: User:Knipptang

Image:Kolkatatemple.jpg *Source*: http://en.wikipedia.org/w/index.php?title=File:Kolkatatemple.jpg *License*: unknown *Contributors*: Original uploader was Nikkul at en.wikipedia

Image:Kolkata Tipu Sultan's Mosque3.jpg *Source*: http://en.wikipedia.org/w/index.php?title=File:Kolkata_Tipu_Sultan's_Mosque3.jpg *License*: unknown *Contributors*: P.K.Niyogi at en.wikipedia

Image:Tagore3.jpg *Source*: http://en.wikipedia.org/w/index.php?title=File:Tagore3.jpg *License*: unknown *Contributors*: Badmachine, Materialscientist, Nagy, Otterathome, Ragib, Roland zh, Saravask, Túrelio, 8 anonymous edits

Image:Swami Vivekananda-1893-09-signed.jpg *Source*: http://en.wikipedia.org/w/index.php?title=File:Swami_Vivekananda-1893-09-signed.jpg *License*: unknown *Contributors*: Original uploader was Dziewa at en.wikipedia

File:394 baul-singers-sml.jpg *Source*: http://en.wikipedia.org/w/index.php?title=File:394_baul-singers-sml.jpg *License*: unknown *Contributors*: Codeispoetry, Encyacht, Mattes, Schimmelreiter, ~Pyb, 1 anonymous edits

File:Dance with Rabindra Sangeet - Kolkata 2011-11-05 6669.JPG *Source*: http://en.wikipedia.org/w/index.php?title=File:Dance_with_Rabindra_Sangeet_-_Kolkata_2011-11-05_6669.JPG *License*: unknown *Contributors*: User:Gangulybiswarup

File:Pitha for Wedding- Pakan, Patishapta, Bharandash.jpg *Source*: http://en.wikipedia.org/w/index.php?title=File:Pitha_for_Wedding-_Pakan,_Patishapta,_Bharandash.jpg *License*: unknown *Contributors*: M. Tawsif Salam

File:Durga Puja celebration.jpg *Source*: http://en.wikipedia.org/w/index.php?title=File:Durga_Puja_celebration.jpg *License*: unknown *Contributors*: User:Dipankan001

File:Mchbuilding.JPG *Source*: http://en.wikipedia.org/w/index.php?title=File:Mchbuilding.JPG *License*: unknown *Contributors*: User:Dwaipayanc

File:Darjeeling from above St. Paul's School.jpg *Source*: http://en.wikipedia.org/w/index.php?title=File:Darjeeling_from_above_St._Paul's_School.jpg *License*: unknown *Contributors*: University of Houston Digital Library

File:IIT KGP Main Building.JPG *Source*: http://en.wikipedia.org/w/index.php?title=File:IIT_KGP_Main_Building.JPG *License*: unknown *Contributors*: Saikat Sarkar

File:Iimc audi.jpg *Source*: http://en.wikipedia.org/w/index.php?title=File:Iimc_audi.jpg *License*: unknown *Contributors*: Rgt18 at en.wikipedia

File:Salt Lake Stadium - Yuva Bharati Krirangan , Kolkata - Calcutta 5.jpg *Source*: http://en.wikipedia.org/w/index.php?title=File:Salt_Lake_Stadium_-_Yuva_Bharati_Krirangan_,_Kolkata_-_Calcutta_5.jpg *License*: unknown *Contributors*: Arindam Ghosh - Rana - Dimpy - lovedimpy

File:Eden Gardens.jpg *Source*: http://en.wikipedia.org/w/index.php?title=File:Eden_Gardens.jpg *License*: unknown *Contributors*: Chippu Abraham from Gurgaon, India

Printed by Books on Demand GmbH, Norderstedt / Germany